How to Rebuild and Modify

Rochester Quadrajet Carburetors

CLIFF RUGGLES

CarTech®

CarTech®

CarTech,® Inc.
6118 Main Streeet
North Branch, MN 55056
Phone: 651-277-1200 or 800-551-4754
Fax: 651-277-1203
www.cartechbooks.com

Edited by: Josh Brown

ISBN 978-1-932494-18-1
Item No. SA113

Written, edited, and designed in the U.S.A.
Printed in China
15

Front Cover:
Barry Rabotnick entered this 427 high-riser engine in the 2009 Engine Masters Challenge. It features a Genesis block, forged-steel crank, set of Carillo H-beam rods, and CNC-ported Blue Thunder heads with 11.5:1 compression. The engine cranks out 670 hp on 91-octane pump gas. (Courtesy of Robert McGaffin)

Back Cover Photos
Top Left:
Most late-model Quadrajets have the external vent. This was discontinued around 1970, proving to be unsatisfactory for the emission standards mandated at that time.
Top Right:
The correct removal procedure for the power piston is to depress and release it several times. The upward spring tension usually pushes the power piston and retainer out of the bore. If this procedure doesn't work, don't bend up on either arm of the power piston. Instead, use a small screwdriver and pry up gently on the centre of the arm as close to the piston as possible. Bent power-piston arms are one of the most common reasons that rebuilt Quadrajet carburetors do not work as intended.
Middle Left:
The tip-in procedure is used to set the APT on the late-model carburetor.
Middle Right:
A number of different accelerator pumps were used. Here is a sample of used pumps removed from different year and model carburetors.
Bottom Left:
A small amount of solder or lead can be melted into the power piston bore to seal off the vacuum supply.
Bottom Right:
Warped base plates and air horns can be trued by sanding these components flat on a bench-mounted belt sander.

DISTRIBUTION BY:

Canada
Login Canada
300 Saulteaux Crescent
Winnipeg, MB, R3J-3T2 Canada
Phone: 800 665 1148 • Fax: 800 665 0103
www.lb.ca

Europe
PGUK
63 Hatton Garden
London EC1N 8LE, England
Phone: 020 7061 1980 • Fax: 020 7242 3725
www.pguk.co.uk

Australia
Renniks Publications Ltd.
3/37-39 Green Street
Banksmeadow, NSW 2109, Australia
Phone: 2 9695 7055 • Fax: 2 9695 7355
www.renniks.com

CONTENTS

PREFACE

The motivation to provide the information contained in this book came from many sources. My friends and fellow enthusiasts encouraged me for many years to make more information available. As fuel injection went into use across the board in 1988, I dismissed putting anything into hard copy and concentrated efforts on building carbs. The Internet made me realize that there are literally thousands of folks in need of assistance and there simply wasn't enough time to help each and every one of them despite my best efforts. The answer was to publish the following material. I hope the information will be beneficial. The information included in this publication comes from countless hours of building, tuning, and testing Quadrajet carburetors. I have included a lot of ideas that I've developed over two decades of working with these carburetors and tried to present it in an easy-to-understand format.

INTRODUCTION

Quadrajet carburetors began showing up on factory production General Motors vehicles in 1965 on Chevrolet 396 engines. Due to ever-tightening emission standards, the auto-making industry was being forced to clean up the amount of pollutants exiting the tailpipes of production vehicles. At the same time, the end of the decade brought with it the peak of the muscle car era. Buyers demanded performance, and the automakers were in a tight race to produce the fastest and most appealing vehicles. Mustangs, Camaros, 442s, GTOs, Roadrunners, and many others showed strong sales and ruled the streets.

These events had the engineers scrambling to produce powerful cars with large-displacement engines, and, at the same time satisfy the emission restrictions mandated by the EPA. Several engine design changes were employed. Chrysler and Pontiac modified the shape of the combustion chambers in many of their engines in 1968. Compression ratios were lowered in 1970, and again dramatically across the board in 1971. Power production fell hand in hand with the drop in compression. Rumors of oil shortages were to follow, and EGR valves and air pumps became standard equipment. Fuel prices began to soar, the reign of the muscle car was at an end.

For General Motors, the Quadrajet carburetor (Q-jet) was a key player during this period. It went through several design changes between 1967 and 1974, and then it was given a complete makeover in 1975–1976. Each change had a purpose, and most of them hand nothing to do with high performance. The competition between the automakers had quickly gone from making the most power and fastest cars to having the least amount of hydrocarbons exiting the tailpipes. The Q-jet survived the battles, while the factory Holley, AFB, AVS, and even the Thermoquads eventually disappeared. The Q-jet made it through, and even after computer engine management systems became standard equipment, the Q-jet stayed in production. By 1988, GM had fuel injection in place across the board, and we saw the very last of the computer-controlled carburetors disappear. Like it or not, our fuel and spark curves were now controlled by electronic components; carburetors became like dinosaurs.

Even though the muscle car era came and went, the interest in high-performance vehicles never died. In fact, it is probably stronger now than ever.

Countless enthusiasts spend hundreds of hours each year building, tuning, racing, and enjoying their high-performance vehicles. The interest in the Quadrajet carburetor has become increasingly strong. A high percentage of hobbyists demand that ALL of the factory equipment be used for the restorations, and nearly as many love to use the factory parts as part of their high-performance vehicle package. The Q-jet carburetor has met considerable resistance when employed as a "high-performance" part. Yet, there is no doubt that they can be made to work. Just attend any IHRA- or NHRA-sanctioned event and watch the Stock and Super Stock cars yank the front wheels into the air and set new records nearly every season. Even so, from the hobbyists' point of view, the Q-jet is simply too complicated and results are simply too unpredictable, even after spending considerable manhours attempting to make one work correctly.

Well, we are going to change all that! In this book, all of the secrets of the Q-jets are revealed. We're going to go way beyond just jet and rod changes. Each system is broken down in great detail. The flow of air and fuel through the carburetor is described, so the reader not only knows where and how to make modifications, but why they are making them. Specific modifications are described and the intended results outlined. We're going to take the guesswork out of correctly setting up your carb for high-performance use. The different models are also discussed, as are factory changes and why they were made. We're going to guide you through building a top-performing carburetor exactly for your application, and most of it can be done with a very minimum investment of tools and equipment. If you ever wanted to know how the Stock and Super Stock racers do it, and how those nearly 4,000-pound factory vehicles that sport a Quadrajet carburetor find their way into the winner's circle at your local track, keep reading, because all the secrets are found in these pages.

The Quadrajet carburetor is in no uncertain terms an amazing creation. Its purpose in life was to provide a highly efficient fuel-metering device using small primary side bores and huge secondaries for outstanding full-throttle performance. Almost anyone who mingles in high-performance circles has at one time or another had an experience with a Q-jet carburetor. Sadly, in many cases these associations have met with less-than-satisfactory results.

After spending over 25 years rebuilding, recalibrating, modifying and testing Q-jet carbs, it's now time for me to make the results of these efforts readily available. We are going to take you step-by-step through the entire Q-jet carb, system-by-system. Different models are discussed, as are parts interchangeability, high-performance modifications, and specific modifications needed for specific applications. The modifications are not only being explained in great detail, but most can be made without going broke from buying a bunch of special tools. Most of the modifications don't even require extensive skills or vast experience with carburetors.

HISTORY

The Rochester Quadrajet was a standard production four-barrel carburetor used by every General Motors division for nearly 20 years. Although it never deviated far from its original basic design, the Quadrajet underwent several modifications over the years, proving its versatility over a wide variety of applications. Some of these changes are covered in this chapter. But those that are most relevant when preparing a Quadrajet for high-performance use are covered in greater detail in later chapters.

It is important to understand the basic Quadrajet design and reasoning behind the changes made throughout its production span. We must first realize that from a practical standpoint, most Quadrajets were never intended to be "high-performance" carburetors. General Motors produced vehicles for a very broad customer base, and each division developed its own engines. Since many different engines were offered in a wide variety of vehicles, nearly every carburetor was calibrated for the specific application. With ever-increasing federal emission standards, manufacturers were forced into making their engines as efficient as possible, which led to future changes. But to better understand the Quadrajet, we must start right from the beginning.

The Rochester Products division of General Motors was located in Rochester, New York, and had built single- and multiple-bore carburetors for GM vehicles since 1949. The AC Delco division, however, issued service information and distributed the carburetors and service components. Rochester's 4-barrel casting

Shown here is a 1966 Quadrajet, carburetor number 7026260. This basic design was typical of very early production carburetors. Carburetors produced in later years would appear similar externally, but many internal changes were made to make them more reliable.

known as the 4G had been one of GM's standard production carburetors from the mid-'50s until the mid-'60s. But as economy and emissions concerns grew, Rochester responded with an entirely new 4-barrel carburetor, the Quadrajet. It boasted small primary bores for maximum throttle response and fuel efficiency, and larger secondary bores to meet the overall flow demands under full-throttle conditions. The Quadrajet was designated the 4M series. This would be the basic casting from which all future variations evolved.

Chevrolet was the first and only GM manufacturer to use the 4M Quadrajet in 1965 on certain V-8 applications. The 4G 4-barrel, however, was used on other Chevrolet V-8 applications, while radical, high-performance applications retained the Holley 4-barrel. Two different Quadrajet product numbers were produced—the 7025200 for 396-cubic-inch engines with automatic transmissions, and the 7025201 for 396-cubic-inch engines with manual transmissions. As the model year progressed, product numbers 7025220 and 7025221, with what apparently revised choke settings, superseded the previous numbers. The product identification number was stamped into a circular disc located on the driver's side of the main body. Identification is discussed later.

The Quadrajet saw expanded use onto a wide variety of 1966 Chevrolet V-8 car and truck engines with displacements that ranged from 327 to 427 cubic inches. It even saw use on select overhead-cam 6-cylinder applications. Buick also selected the Quadrajet for its 400 and 425

cubic-inch engines in 1966. Oldsmobile followed suit. Cadillac and Pontiac were not as quick to change, but by 1967 they too had begun installing the Quadrajet on most of their 4-barrel applications. With its growing popularity and efficient design, the Quadrajet was the most widely used production 4-barrel carburetor on GM-produced engines during that time. And though other carburetors such as the Rochester 4G, the Carter AFB, and the Holley 4-barrel were used on certain engines, the Quadrajet quickly became a benchmark for other 4-barrel carburetor manufacturers.

One of the first major changes to the Quadrajet design was making it available in both front- and side-inlet models, shown here. The front inlet is shown on the right; side inlet is on the left.

The 1968 model year marked the first nationwide federal standards for specific pollutants. This not only brought on internal engine changes, but many GM engines received components that controlled the spark advance and fuel curves throughout the entire operating range. And because of its efficiency and excellent balance of economy and performance, the 4M Quadrajet became the standard production 4-barrel carburetor for all GM manufacturers including GMC, Checker Cab, and the

Marine division. The Rochester 4G was phased out as the Quadrajet replaced everything except the Holley 4-barrel, which continued on certain Chevrolet applications through 1972.

With the Quadrajet used extensively by all GM divisions, each manufacturer required certain external characteristics that matched its under-hood routings. The most common difference many hobbyists are most familiar with is the forward-facing fuel inlet as opposed to those that are side facing. And since the Quadrajet was used on such a wide variety of engine sizes from each respective automobile manufacturer, separate fuel-metering circuits and a variety of vacuum sources were also required and each carburetor was given a specific casting number.

Demand from the divisions had increased to the point that Rochester could not meet the volume requirements. To prevent production delays, GM approached the Carter Corporation about producing Quadrajets in addition to Rochester. The Carter-built units are nearly identical to the Rochester-produced Quadrajet—the only differentiating characteristic is "Quadrajet by Carter" cast in place of "Quadrajet by Rochester" on the side of the main body. Since Carter produced Quadrajets well into the 1970s, they are not necessarily uncommon. Either would make an acceptable starting point for performance modifications.

Not unlike anything designed from a clean slate, the early Quadrajets were plagued by several small problems, such as the plunger-style fuel valve and the

The carburetor on the left is a Rochester 4GC, used widely by General Motors until the Quadrajet (shown on the right) was introduced.

In the late 1960s, Carter was contracted to produce Quadrajet carburetors to help keep up with production. The Carter units were manufactured under contract and clearly state "MFD. By Carter Carburetor for GMC" as seen in the bottom casting.

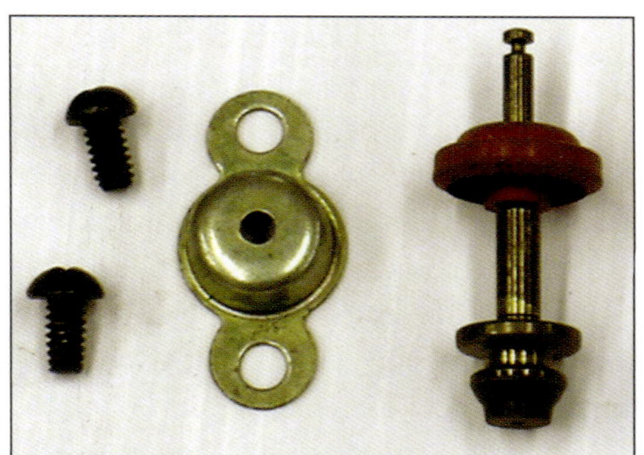

Very early Quadrajet carburetors used a plunger-style fuel-valve assembly. These were prone to trouble and disappeared in 1967.

secondary-air-valve dash pot assemblies. But they were improved upon and increased overall function and reliability in subsequent years.

Many of these are discussed later in the book. But the first notable changes occurred for the 1969 model year. A smaller float with a relocated fulcrum increased float-bowl volume and provided more pressure on the fuel needle to better control the fuel level. Though these types of changes were typically implemented in all the Quadrajets produced, the Oldsmobile and marine applications retained the earlier-style float for several years.

Another series of changes occurred in 1971. Plastic caps were placed over the idle mixture screws on the base plate of certain Quadrajets. These were used to limit travel and prevent grossly inaccurate adjustments, which could affect emissions. Internal metering changes were also required, with General Motors implementing a maximum compression ratio of just 8.5:1, which applied to each division for the 1971 model year. This was not only an attempt at reducing unburned hydrocarbon and oxides of nitrogen emissions, but also allowed the engines to operate on low-lead or unleaded fuels. The federal government had proven that lead particulates emitted from automobile tailpipes were finding their way into environmental areas such as water streams and farming soil. These negative effects on the environment and humans prompted governing bodies to impose limits on the amount of lead present in fuel. Eventually it was removed entirely.

Leaded gasoline had been available for several decades prior. The lead was added in an attempt to increase the overall octane rating of the fuel but was later

Moving the float fulcrum forward for 1969 (carburetor on right) models was one of the first major design changes for the Quadrajet. The early-style float arrangement continued in production well into the 1970s for various applications.

Many early models featured plastic caps over the idle-mixture screws to discourage tampering. It's quite rare to see any in place these days. They were one of the first items removed from the carburetor during basic engine tune-ups.

In 1971 Pontiac used a special version of the Quadrajet lacking the outer booster rings. The idea was to provide additional airflow, but these carburetors lacked the part-throttle efficiency of standard units and were difficult to get through emission testing. They were discontinued by the 1972 model year.

found to also improve the life of the exhaust valve seats. Since lead does not burn, it typically left behind a powdery residue, which coated the entire combustion chamber, including the exhaust valve and exhaust valve seat as the gasses exited. As lead was removed, the extreme heat of the exhaust gasses passing over the valve seat and the rotating action of the valve itself generally wore untreated cast- iron seats more quickly, especially with an engine that was frequently under heavy load. It was determined that the lead acted as a "lubricant," preventing direct contact between the valve seat and the valve and reducing the wear process.

Shortly after lead restrictions were enacted, many GM divisions began taking measures to reduce wear by induction-hardening the exhaust-valve seats on their cylinder heads. This process incorporates heating the valve seat to approximately 1800 degrees F for a short time with a coiled wire fixture. As the seat cooled, the result was an electrically inducted, hardened seat that extended about .050-inch below the surface. This greatly increased exhaust valve seat life and improved engine performance. The compression ratio loss took its toll, but

performance was still strong from all the GM manufacturers. While the others may have approached improving lower-compression performance in a different manner, Pontiac and Buick both compensated by addressing the Quadrajet.

Pontiac modified the basic 4M Quadrajet casting by removing the outer velocity boosters that typically surround the discharge nozzles located in the center of the primary bore in the main body. A former Pontiac engineer noted that this significantly increased the airflow capacity of the carburetor and improved performance. But because of the weaker vacuum signal from the missing boosters, throttle response was degraded slightly, making them more difficult to tune and pass off-idle and low-speed emissions standards. The design was used for just one year. Five different casting numbers exist from 1971 Pontiac applications. They include 70-1267, 7041268, 7041270, and 7041273, which are specific to 455 HO engines, and 7041263, which was used on 400 and 455ci engines with manual transmission. These are extremely rare and highly coveted by the Pontiac crowd. These particular carburetors were more than just stock castings with the outer

The 1971 Pontiac HO carburetors, lacking the outer booster rings, were actually special castings. Shown here is the bottom of the single booster extending well into the main primary bore.

booster rings removed. The center booster was completely redesigned to extend almost 1 inch further down into the main venturi area.

Buick improved the performance of its large-displacement engines by increasing the airflow capacity of its 1971 Quadrajets. Rather than expanding on Pontiac's approach, they simply enlarged the primary bore in the main body. Typically 1½ inches in diameter, Buick's modifications increased bore diameter to 1⅞ inches.

Generally referred to by airflow capacity of "approximately 800 cfm," Buick continued using this design through 1974. Its advantage of additional airflow improved performance, but unlike Pontiac's approach, Buick's Quadrajet retained the velocity boosters for maximum throttle response and improved emissions. These early Buick carbs are an excellent starting point for a high-performance carburetor and are the only large CFM castings produced with divorced-style chokes.

Plastic limiter caps were installed onto the idle mixture screws of all Rochester carburetors for 1972. That year saw another major step in the reduction of exhaust emissions with the introduction of an exhaust gas recirculation (EGR) valve onto certain Buick applications. With federal oxides of nitrogen (NOx) standards implemented for 1973, all GM divisions used the EGR system extensively that year. A typical system consisted of a vacuum-actuated valve mounted on the intake manifold. The valve allowed exhaust gas passing through the exhaust crossover to circulate back into the intake plenum, where it would intermix with the incoming mixture of fuel and air. Since the production of NOx is directly related to combustion temperatures, the exhaust gas actually cooled the combustion chamber, reducing the amount of NOx under light load.

To maximize the effects of the EGR system, the primary circuits of Quadrajets destined for EGR applications was slightly leaner. Since exhaust gas is inert, it does not burn under normal combustion, and in theory it occupies volume normally occupied by the combustible air/fuel mixture. And with less combustible air in the chamber, less fuel is required. With less combustible mixture, flame intensity cannot be as great, which reduces combustion temperature and the production of NOx.

The vacuum source for the EGR valve was a specific vacuum port located in any number of places on the Quadrajet main body or base plate. Since EGR could cause a slight hesitation just off idle or a lean surge when decelerating, the EGR valve was only active at part-throttle cruise. The EGR-specific vacuum port reflected these requirements. A special timed port provided varying amounts of vacuum that increased proportionally with throttle angle to meter the varying amounts of exhaust gas that recirculated into the intake manifold. Because of its unique vacuum signal, the EGR vacuum port makes neither an acceptable ported nor manifold vacuum source. The high source location in the base plate does not provide a strong enough signal right off idle to effectively operate the distributor vacuum advance.

Pontiac engineers used a variation of Buick's large-bore carburetor on its potent Super Duty 455 in '73–'74. These high-performance carburetors are among the best for modification. They generally

Here's a great salvage-yard find, a 1971 Buick carburetor. These carburetors were the first large primary bore castings, often referred to as "800 cfm" carburetors. They are easily identified by the slashed front vent tube and long curved arm on the throttle linkage.

Many Quadrajets have special vacuum source locations specifically designed for EGR valve operation. Shown here is a 1974 Pontiac base plate noting the EGR port.

have larger internal fuel passages and generous jetting. The large-bore Pontiac carburetors have a carburetor-mounted thermostatic coil assembly on the main body. Buick's model received a manifold-mounted thermostatic choke coil. Both of these Quadrajets are considered the best of the 4M-series for high-performance use. Since the Super Duty carburetors used the hot-air-style choke, the choke pull-off was moved to a separate bracket bolted to the front of the carburetor.

When hot-air chokes were added to the main casting, the choke pull-off received its own bracket as noted on this 1973 Pontiac Super Duty Carburetor.

The 4M casting was redesigned for 1975 and is referred to in Delco literature as the "Modified Quadrajet" and designated the M4M series. Most retained the larger 1$\frac{3}{2}$-inch primary bore like the Buick and Super Duty Pontiac 4M, though some castings bound for low-performance, small-displacement applications retained the 1$\frac{3}{2}$-inch bore. Regardless, the new modified castings also included several design changes. The most notable was a much smaller float design. This increased fuel-bowl capacity while maintaining sufficient pressure on the needle/seat assembly to control fuel level. Other changes included an adjustable part throttle (APT) system located in the right front corner of the main body. A brass cap in the air horn was

In 1975 the Quadrajet got another float fulcrum modification as shown on this 1975 unit.

located directly above an adjustment screw and was used to fine-tune part-throttle air/fuel mixture ratios.

In addition to APT, many 1975 model-year Quadrajets received an additional power piston and single jet/rod assembly. The intention was to provide closer control over the air/fuel ratios while operating on the primary circuit. However, this was not as effective as initially desired and was removed from most carburetors by the 1976 model year. The carburetors using the auxiliary

Some 1975–'76 models used an auxiliary power piston with a single metering rod and pressed-in jet. This system worked in conjunction with the APT system and the main jets and metering rods.

power piston assembly are somewhat difficult to set up for performance use. Although they can be modified with some additional effort by completely blocking off the auxiliary power piston and single pressed-in jet, it is usually better to simply obtain a later-style carb that didn't use this system in the first place.

The 1976 castings are the first of the more desirable M4M Quadrajets. They contain many features that make them an excellent starting point for high-performance use. With an external appearance identical to the 1975 castings, they too use the small float design. The APT system, however, was relocated next to the primary power piston in the void left from the auxiliary power piston that was used in 1975. A black plastic cup filled the cavity left from the 75-type APT. The power piston was equipped with a small pin that contacted the APT adjustment screw, which could be adjusted externally. This new-type APT system allowed for full control over part-throttle air/fuel ratios by simply repositioning the tapered portion of primary rod in the primary jet. Because of this flexibility, these carburetors are much easier to tune than earlier units.

The Quadrajet continued for the rest of the 1970s without any significant changes. In 1979, however, the plastic limiting caps on the idle mixture screws were eliminated. Hardened steel caps pressed into the base plate took their place. This made any additional tuning impossible and prevented any adjustment that could affect tailpipe emissions. Several methods of removing these plugs have evolved over the years, but the most effective method was printed in AC Delco service literature. Centerpunch and a sharp hacksaw blade are used to score and cut the front edge of the base plate. The main drawback is the rough cutting marks left behind on those that have had the steel plugs removed. Once removed, however, these base plates are as good as any other.

A plastic cup fills the location once used by the short-lived 1975–'76 APT models.

A pretty rare sight: a late-model Quadrajet that still has caps over the idle-mixture screws. Most were removed by technicians due to customer complaints during the warranty period!

This later APT system replaced the single pressed-in jet and metering rod used in the 1975 and some 1976 models. The new system had a provision to raise and lower the power piston. It also provided fine metering control with tapered second-step metering rods, or when two-step rods were raised to the tapered section between steps.

The start of the 1980 model year was virtually a carryover from the previous year. One noticeable change was the choke system. The screws that typically fastened the thermostatic choke coil to its housing were replaced by rivets. This was another attempt for better control of emissions by removing the chance of an incorrect choke valve adjustment.

Another factory modification to resist tampering was to install rivets over the choke cover screws in lieu of #8-32 screws.

One of the most significant changes made to the Quadrajet since the inception of its original design occurred during the 1980 model year. With the introduction of Computer Command Control (CCC) on several GM vehicles, the Quadrajet was redesigned to accept electronic internal components, which offered more precise control over main-metering-system-reducing emissions and improving economy at part throttle. This latest version of the basic Quadrajet was given a model designation of E4M, which signified an Electronic Controlled 4M carburetor. The secondary circuit, however, was not affected and operated conventionally.

The heart of the CCC system was an Electronic Control Module (ECM), which monitored several engine parameters under normal operating conditions. An oxygen sensor in the exhaust stream and an electronic throttle position sensor

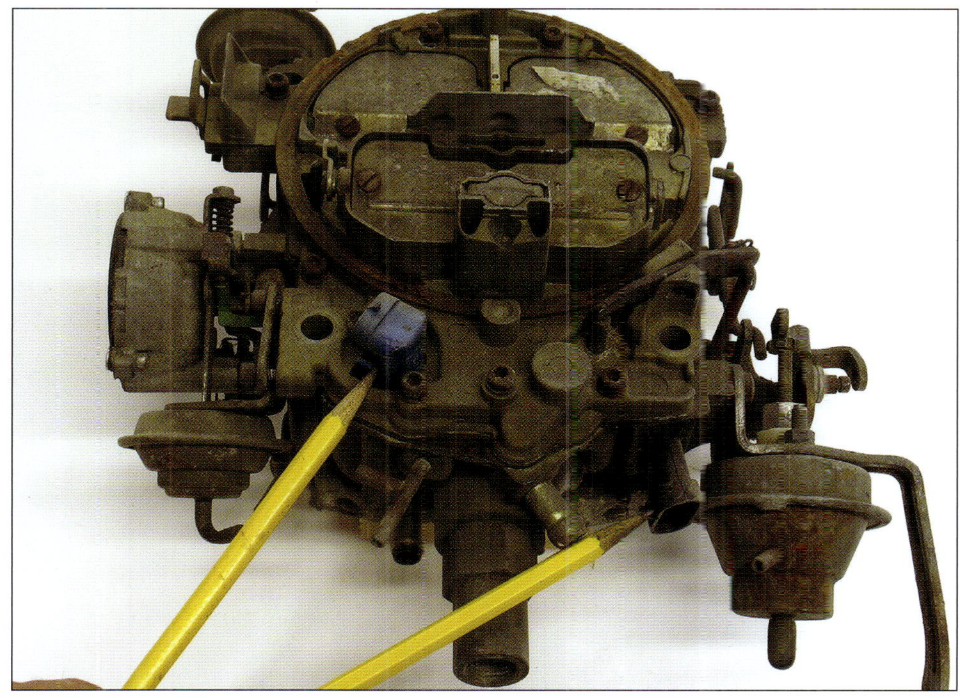

Quadrajets started showing up in the 1980s with electronic mixture-control solenoids (left, blue) and throttle-position sensors.

within the main body of the carburetor sent electronic signals to the ECM. Based on the input, the ECM sent varying signals to an electronic mixture-control solenoid within the carburetor, which controlled the depth of the primary metering rods in their jets, richening or leaning the mixture based on engine load. This procedure replaced the vacuum-sensitive primary power piston and APT system in the conventional M4M series Quadrajet. A variation of this design retained the conventional power piston but contained an electronic mixture-control solenoid that controlled part-throttle mixture with a '75-type APT.

The E4M Quadrajets were not necessarily problematic, but they can be very difficult to calibrate for performance applications that do not use an ECM and all related equipment. The primary bore was typically 1³⁄₃₂ inches and they were calibrated for low-performance, small-displacement engines where economy and emissions were of the utmost concern. And because of this, finding larger primary jets and smaller primary rods for performance applications can be very difficult. They can make an excellent donor to salvage parts from since so many parts interchange with other 4M and M4M Quadrajets. However, once the electronic equipment is removed

and modified correctly, the E4Ms make excellent performance carburetors.

The M4M Quadrajet was used extensively by all GM divisions. Remarkably, they were even installed on smaller-displacement engines such Chevrolet's 305ci and Pontiac's 301ci. This just proves how versatile the Quadrajet is—the same basic casting can be found on a wide array of GM engines with displacement ranging from as little as the 300ci engines from Chevrolet, Oldsmobile, and Pontiac, to Cadillac's enormous 500ci mill. Regardless of its original application, we hope to show how any Quadrajet can be modified to fit the needs of another application.

Here is another 1980s-era model with a dual-capacity accelerator-pump solenoid to reduce the accelerator-pump shot once the engine has reached operating temperature.

Carburetor Designation/Identification

Although many castings are referred to by specific characteristics such as airflow capacity or choke type, Rochester actually gave each Quadrajet style different designations indicating body type and choke style. The first design series was designated 4M and can be found on all GM vehicles through 1974 and on GMC and Chevy trucks through 1977. When Rochester redesigned the Quadrajet in 1975, they referred to it in factory literature as the "Modified Quadrajet" or "Mod-Quad," and designated it M4M. When computer-operated electronic internal components were added around 1980, Rochester designated the Electronic Quadrajet the E4M.

The last letter in the series designation generally indicates choke type. The letter "V" indicates an automatic choke where the thermostatic choke coil is fastened to the exhaust crossover on the intake manifold. Commonly referred to as a "divorced" choke, a small rod connects the coil to the mechanical linkage on the passenger side of the main body. The letter "C" indicates a carburetor-mounted automatic choke. In these instances, the thermostatic coil, which is mounted on the passenger side of the main body, uses heated vacuum and is easily recognizable by its large, black adjustment disc. They are generally referred to as a "hot air" style. The letter "E" indicates an electronic choke coil. They too, are recognizable by a large black disc, but they operate off of a 12-volt source as opposed to vacuum. And those Quadrajets without any letters following the 4M and M4M designation had a manual choke.

The list below identifies the different types of Quadrajet designations that appear in AC Delco printed information.

4M:	Quadrajet with manual choke
4MC:	Quadrajet with carburetor-mounted automaticchoke
4MV:	Quadrajet with divorced automatic choke
4ME:	Quadrajet with carburetor-mounted electric choke
M4MC:	"Modified" Quadrajet with carburetor-mounted automatic choke
M4ME:	"Modified" Quadrajet with carburetor-mounted electric choke
M4MEA:	"Modified" Quadrajet with carburetor-mounted electric choke and "Altitude" calibration
E4MC:	"Electronic" Quadrajet with carburetor-mounted automatic choke
E4ME:	"Electronic" Quadrajet with carburetor-mounted electric choke

Most carburetors produced in and after about 1975 have the choke housing mounted on the carburetor. The thermostatic coil is activated by hot air from the intake manifold exhaust crossover system via a heated tube. Later, electric chokes started showing up in certain applications to replace the hot-air-style chokes.

Carburetor Numbering

Through the years of production, the manufacturer used numbers for carburetor identification. Very early Quadrajets had tags installed into a cast round area on the driver's side of the main body.

Electric Chokes

An electric choke can significantly improve cold-weather operation over other original choke styles.

Any carburetor originally fitted with a hot-air thermostatic choke coil can easily be converted to the later GM electric choke. However, to prevent a vacuum leak, the hot-air choke uses a gasket between the choke coil and housing. Since the housing acts as a ground for the electric choke, this gasket must be omitted and the vacuum sources plugged when converting. New electric coils should be available from your local parts store or even a junkyard. All that's needed for proper choke operation is a 12-volt power source that is activated when the ignition key is in the "ON" position. A good power source location is at the windshield wiper motor. It is usually mounted on the firewall behind the engine and has a wire that is activated when the key is switched to the "ON" or "RUN" position.

Any carburetor originally equipped with a hot-air-style choke can be converted to an electric choke in minutes. The vacuum supply to the housing must be blocked off and the gasket under the choke discarded because the unit uses the housing as a ground.

Early Quadrajets used a circular metal tag to convey the carburetor number.

The 6th number indicates the division application. "0," "1," or "2" is Chevrolet; "3" is Cadillac; "4" is Buick; "5" is Oldsmobile; "6" and "7" are Pontiac; and "8" is Checker and Marine. The last number indicates the exact application. For the most part, even numbers were used on automatic transmissions, and odd numbers on manual transmissions, but there were numerous exceptions.

For carburetors produced from 1970 to 1974, the first three numbers are "704" followed by the same sequence of numbers as the "702" series of carburetors. (Some 704 carburetors were produced early in the 1975 model run; later units start with "1705.")

Staring in the 1975 model year, a "1" was placed in front of the "7" followed by "705." The "1705" series of carburetors would run until 1979.

In and after about 1968, stamped numbers were used for identification. The large flat area on the main body just above the secondary throttle shaft has a series of numbers stamped into the casting. In most cases these numbers can be used to identify the year and application the carburetor was originally used on.

Carburetors produced until 1969 begin with a 702 followed by four additional numbers. Some Carter-produced models lack the "70." The 4th number indicates the last number of the application year. The 5th number is either a "1" or "2," indicating federal emissions, or a "4" or "5" denoting California emissions.

The 1705XXXX-series carburetors were produced from 1976 through 1979. The 17057266 carburetor is a 1977 model-year Pontiac carburetor built on the 348th day of 1976.

Beginning in 1980, the first four numbers were changed to "1708," and this number series would continue until the last carburetors were produced in the late 1980s.

Plant codes may also be located with the stamped numbers, usually two or three letters and the date that the carburetor was produced. Most Rochester-

In 1968 both Rochester and Carter started stamping the carburetor number onto the main casting just above the secondary linkage.

704 1 2 4 0

Casting Numbers-
702xxxx- 65-69 Federal
703xxxx- 65-69 California and Altitude applications
704xxxx- 70-75

- 4th digit denotes the last numeral of the application year.

- 5th digit, typically either "1" or "2" denotes if the original application is federal while "4" and "5" denote California applications.

- 6th digit typically indicates division application (0,1,2-Chevy; 3-Cadillac; 4-Buick; 5-Olds; 6,7-Pontiac; 8- Checker and Marine).

- 7th digit typically indicates transmissions usage; even numbers are many times automatic trans applications while odd numbers are manual trans but this is not always true.

1705 6 2 6 3

Casting Numbers-
1705xxxx- 76-79
1708xxxx- 80-up

- 5th digit denotes the last numeral of the application year.

- 6th digit, typically either "1" or "2" denotes if the original application is federal while "4" and "5" denote California applications.

- 7th digit typically indicates division application (0,1,2-Chevy; 3-Cadillac; 4-Buick; 5-Olds; 6,7-Pontiac; 8- Checker and Marine).

- 8th digit typically indicates transmissions usage; even numbers are many times automatic trans applications while odd numbers are manual trans but this is not always true.

Designations-
- 4G- Rochester 4bbl with manual choke
- 4GC- Rochester 4bbl with automatic choke
- 4M- Quadrajet with manual choke
- 4MC- Quadrajet with automatic choke (carb mounted)
- 4MV- Quadrajet with automatic choke (divorced)
- 4ME- Quadrajet with electric choke
- M4MC- "Modified" Quadrajet with automatic choke (75-up style)
- M4ME- "Modified" Quadrajet with electric choke (75-up style)
- M4MEA- "Modified" Quadrajet with electric choke "Altitude" (75-up style)
- E4MC- "Electric" Quadrajet with automatic choke
- E4ME- "Electric" Quadrajet with electric choke

built carburetors used a four-number stamp indicating the Julian date. The first three numbers are the day of the year the carburetor was built; the last number is the year. For example: carburetor number 17058201 followed by a Julian date of 1238 would have been built the 123rd day of 1978. One must keep in mind when dating carburetors that the year of production may not always be the same year as indicated by the carburetor number. Quite a few service replacement carburetors were produced. It is quite common to find much later production dates than what is indicated by the carburetor number.

In addition, Carter carburetors produced in the late 1960s may have been

The Julian date on this carburetor shows that it was produced several years after the model-year run. These service replacement carburetors were often used to replace troubled units under warranty. Service replacement carburetors were available for many years after the original production dates.

In lieu of using Julian dates, Carter carburetors, both originals and service replacements, used a Letter/Number for date coding. The "H8" stamped on this 1969 Chevrolet carburetor indicates that it was produced in August of 1968.

used in a different model year than the number shown on the carburetor. This was most common in the 1968–1969 models. Rochester was not able to keep up with production. Many Carter Quadrajets produced in these years may have been carried over to the next model year. This information is most important for restorers. It could be correct to have a fully restored 1969 Pontiac GTO using a Carter carburetor produced in 1968.

It is important to note that there are quite a few exceptions to the basic numbering we have described above. For the most part the information holds true as we search swap meets and salvage yards for good cores. Regardless of what number is depicted on the carburetor, one should also give the carburetor a close visual inspection. It is quite common these days to find carburetors that have been re-stamped and others that are missing the numbers entirely. With rare high-performance carburetors bringing in excess of $2,000, there are always those who are going to re-stamp low-performance units.

With the value of original high-performance HO and Ram Air carburetors at an all-time high, re-stamping standard units to sell for high profits is commonplace. The quality of the work varies greatly; this unit is an obvious re-stamp as it contains no date or plant code, in addition to not having the numbers lined up very well!

HOW QUADRAJETS WORK

The first approach to working with most anything is to become familiar with how it works. With the Quadrajet carburetor in particular, this is of the utmost importance. It is not only important to understand the basics of carburetor function, but also how each system within the carburetor works. The first step is to become familiar with the various parts of the carburetor, including the different systems and components within them. Next, we must get a basic understanding of how fuel and air flow through these systems over the range of engine operation. Armed with accurate information, one can make the needed modifications for performance improvements with great confidence.

The Quadrajet, as with any other factory-delivered carburetor, was set up specifically for a single application. The engineers designed and calibrated the entire vehicle, and each component, including the carburetor, to work best within given parameters. In most cases only a correct rebuilding is all that is needed to get the carburetor to work well in all areas, provided the engine and all associated emission equipment are still in place and operational. However, we all know this is quite rare. In the vast majority of cases, the well-meaning engine builder/tuner makes significant changes to the engine in their quest for improved performance. Once one decides to employ a non-factory intake, exhaust, and camshaft, or even to alter the engine's static compression ratio, we now become our own design engineer. The typical result is that all or most of the factory-installed equipment is not up to the task of correctly supplying our engine with the ideal settings over the entire engine speed and load range. Modifications to the different systems within the carburetor now become mandatory if one is to extract the full performance potential from the engine in all areas. This would be an easy task if

An 850cfm custom Quadrajet gets ready to feed the air/fuel to a stout 550 hp pump-gas street engine on Kauffman Racing Equipment's dyno.

the engine operated at a fixed speed and throttle position, but they do not. Specific modifications to nearly every part of the carburetor are typically required. The modifications outlined in this publication are for the most part easy to perform. The hard part is to know where to perform them and why. Read on, because understanding the basic systems can build confidence that your Quadrajet carburetor performs to expectations.

In order to fully understand carburetor operation, one must have a basic understanding of engine function. During various driving situations, the engine operates over a rather broad revolution per minute (RPM) range. The load placed on the engine also varies considerably. The carburetor is asked to deliver the correct amounts of air and fuel in the correct ratio over the entire engine speed and load range.

An engine is defined as a machine that converts heat into mechanical energy. An internal combustion engine is nothing more than an air pump that uses a piston (or pistons) that moves up and down within a cylinder or confined space. The constant up and down movement continually increases and decreases the volume in the cylinder. This movement, combined with intake and exhaust valves, creates the necessary processes that become the controlling forces of engine operation. The vacuum created during the intake or downward stroke of the piston while the intake valve is open causes air to be pulled into the cylinder. This air must pass through the carburetor before entering the intake manifold. The duty of the carburetor is to ensure that the appropriate amounts of fuel are present in the air at all times for sufficient combustion. As we go through the various carburetor systems, one quickly sees that this is a tall order.

Theoretically, the perfect air-to-fuel ratio (A/F) for ideal combustion within the cylinder in pounds of air to fuel is about 15:1. During normal engine operation, the air-to-fuel ratio, or A/F, varies from about 20:1 to about 8:1. Economy ratios are typically in the 14:1 to 17:1 range and power mixtures several ratios richer. At a glance, the task of the carburetor to deliver an ideal A/F of 15:1 seems like an easy function to perform. One would think that all the engineer had to do was to set every carburetor up with an exact calibration that would deliver the ideal A/F ratio. However, considering the obstacles present during engine operation at various speeds and load, perfect conditions seldom exist. There are always varying amounts of exhaust gases remaining in the cylinder(s) during the intake stroke that dilute the intake charge. The air and fuel is seldom perfectly mixed as it enters the cylinder, and the quality of the intake air can vary considerably. In addition, the intake manifold does not deliver equal air/fuel mixtures to all cylinders. Also, changes in the fuel discharge volume from the carburetor do not always result in equal changes in the mixture delivered to the combustion chamber.

To further compound the problems, the engine builder often employs parts designed to increase engine power, specifically a camshaft sporting larger specifications than the factory cam. The longer overlap cycle (both valves open at the same time as the piston passes over top dead center between the exhaust and intake strokes) further dilutes the combined air/fuel mixture that ends up in the combustion space during the compression stroke. Without a proportional increase in the static compression ratio of the engine, we now have an engine

RPM	TQ	HP		RPM	TQ	HP	
4500	528.1	452.5	96.0	4500	540.1	462.8	96.0
4600	523.0	458.1	99.8	4600	536.6	470.0	99.8
4700	505.3	452.2	103.7	4700	525.8	470.5	103.7
4800	498.7	455.8	107.8	4800	516.5	472.0	107.8
4900	503.0	469.3	111.8	4900	510.8	476.6	111.8
5000	492.6	469.0	116.1	5000	510.4	485.9	116.1
5100	474.5	460.8	121.2	5100	492.8	478.5	121.2
5200	461.1	456.5	126.4	5200	483.1	478.3	126.4
5300	457.6	461.8	131.9	5300	470.7	475.0	131.9
5400	447.3	459.9	137.4	5400	464.1	477.2	137.4
5500	435.0	455.5	142.9	5500	451.7	473.0	142.9

The dyno sheet shows back-to-back test runs from a 455ci street engine on Kauffman Racing Equipment dyno facility in Glenmont, Ohio. Part of the testing included intake and carburetor changes. A custom 850cfm 1976 Q-jet was used for the right dyno chart, and a 1970 Ram Air Pontiac 750 CFM carburetor on the left. Both carburetors were carefully calibrated for the dyno runs to make the results as accurate as possible. (Kauffman Racing Equipment)

that is much more difficult to keep running at low engine speeds. This results in uneven firing of the cylinders, low engine vacuum at idle, and that "lope" from the exhaust pipes that many of us have come to love. Unfortunately, most factory carburetors fall way short in being able to deliver the additional fuel required at low engine speeds when the vacuum is relatively low. They simply weren't calibrated for the new engine parameters. We now have to either replace them with a carburetor that has more generous capabilities or make specific modifications so that our existing carburetor is up to the task.

When outlining any carburetor modifications we must also consider that engines vary dramatically in size or displacement. Displacement is usually listed as cubic inch displacement (CID), cubic centimeters (CC), or liters. For comparison, a 400ci engine is approximately 6.6 liters or 6,500cc. Engines not only vary dramatically in size, but also in compression ratio and power output. The Pontiac division of General Motors, for example, delivered its basic engine block through the years of production from 267ci to 455ci. The compression ratio of these engines could have been as low as 7.7:1 to as high as 10.75:1, and the power produced from them varied from just less than 200 to nearly 400 hp!

In early years, well before any significant emission standards were imposed, carburetors were often matched very closely to the size of the engine and expected power potential. Carburetors are typically rated in their ability to flow air in cubic feet per minute (CFM). There are various formulas in place for determining the ideal CFM needed for any engine based on its displacement and RPM potential. A "rough" racer's formula is to simply double the CID of the engine. Although a crude measurement, it has proven to be accurate enough for basic carburetor sizing. What we have seen here from extensive engine dyno testing is

Formula for determining carburetor airflow requirements in Cubic Feet per Minute:		
Street carburetor CFM	=	.85 x RPM x CID / 3456
Street/Strip carburetor CFM	=	RPM x CID / 3456
Racing carburetor CFM	=	1.1 x RPM x CID / 3456
CFM	=	Cubic feet per minute
RPM	=	Engine revolutions per minute
3456	=	Constant for conversion factors
.85, 1, and 1.1	=	Approximate volumetric efficiencies for each type of engine

that smaller carburetors can hurt peak power, but often do not greatly affect overall vehicle performance. For example, we prepared a 455 Pontiac street engine that made 485.9 hp topped with an 850 cfm carburetor. The engine made peak power at 5,000 rpm. According to one well-used formula (see chart) to determine the "ideal" CFM required for this engine, the engine only needed about 644 cfm. Even so, a back-to-back pull was made with a smaller 750 cfm carburetor and the engine made 16.6 less hp!

All carburetors operate on the basic principle of pressure difference. Any pressure less than atmospheric pressure, or about 14.7 pounds per square inch (PSI), is considered a vacuum or low-pressure area. By creating a vacuum or low-pressure area, it becomes possible to cause fluids and air to move or flow from the high-pressure to the low-pressure area. To help increase the pressure drop and improve fuel flow, carburetors use an hourglass-shaped restriction. This device is called a "venturi." Air coming down through the carburetor that encounters the venturi increases in velocity as it passes the narrower area. This creates a low-pressure area and increases the "pull" from the main fuel nozzle. Venturi is designed with a specific size, length, and curvature to be most effective within the design parameters of the carburetor. One must consider that a small venturi increases air velocity at any given CFM or airflow. This can be beneficial for precise fuel delivery and atomization, but may restrict high-speed engine operation.

A large venturi would be more advantageous for high-speed engine operation, but is typically less effective at low engine speeds. The production venturi size is usually a compromise to provide adequate low- and high-speed engine operation. The Quadrajet uses a rather conservative primary side bore and venturi size to be most effective at low and moderate engine speeds. It is complemented by huge secondaries for sufficient airflow at high RPM.

The internal force that causes a pressure drop and movement of air into the internal combustion engine is created by the downward movement of the piston during the intake stroke. The air that enters the engine flows through the carburetor, then through the intake manifold and intake ports in the cylinder head(s). Since the engine operates at various speeds and loads, the amount of air and the pressure drop or vacuum created by the engine varies dramatically. This requires the carburetor to have several different systems to ensure accurate fuel delivery over the engine speed/load range.

The following pages take you through these systems on the Quadrajet carburetor. Understanding the basic function helps us understand where, how, and why to make changes outlined in Chapter 6.

Main Components

The Quadrajet carburetor has three basic components: the main fuel bowl, the air horn, and the throttle body or base plate. We have chosen to call the

A late-style (1975-up) air horn. Please see Appendix B for a guide to the basic components.

throttle body a base plate for the purpose of this publication. This is done solely to eliminate confusion, as the term throttle body, when used these days, refers to part of a fuel-injection system. The outward appearance of the Quadrajet is very similar through the years of production. A major design change occurred in 1975. Most 1975 and later factory units are noticeably different in appearance and can be quickly identified by locating the round boss in the left front corner of the air horn (as viewed from the front).

Regardless of the year of production or application the carburetors would have been used on, the Quadrajet continued to use the same basic systems, which are described in detail in the following pages. When applicable, minor changes are noted. Modifications to these systems are outlined in Chapter 6.

Venting

To control the carburetor bowl pressure, venting becomes necessary. Quadrajet carburetors used both internal and external venting and combinations of both through the years of production. Early designs used external venting in the front of the air horn and allowed fuel vapors a path directly into the atmosphere. This proved to be unsatisfactory to control evaporative emissions and was dropped from the design by about 1970. Due to safety regulations, Quadrajets used in marine applications did not use any external venting.

This main body is part number 1705XXXX, a 1978 Oldsmobile unit. Please see Appendix B for a guide to the basic components.

This photo reflects the base plate. Please see Appendix B for a guide to the basic components.

It is also important to know that air-cleaner design greatly affects the location of the vent tube. These account for the variations in size, angle, shape, and location of vent tubes found on Quadrajets used on various applications. Vent location is also influenced by the fact that the fuel within the carburetor is bouncing around on every turn, acceleration, stop, start, and rough road. It is similar to running up or down a hill carrying a glass of water. The vent must not be located so that fuel is slopped out during these conditions. In addition to well-located venting, the Quadrajet used internal baffles to aid in fuel control. This has given them a good reputation for use in extreme applications, such as off-roading and hill climbing.

Most late factory Quadrajets have the external vent. This was discontinued around 1970, proving unsatisfactory for the emission standards mandated at that time.

Starting in 1971 and used through at least 1979, Pontiac carburetors use the large front vent as shown on the carburetor on the right. The carburetor on the left is a 1970 Pontiac carburetor showing the typical front vent location from that period.

Fuel Inlet

Through the years of production, the Quadrajet was offered with two different fuel inlet locations. For the most part, Chevrolet carburetors have a side fuel inlet. Most Buick, Oldsmobile and Pontiac units have a front inlet. There are exceptions to this basic rule for nearly every division of GM, but for the most part it holds true through the years of production. Several different-length filter housings were used along with both short and long fuel filters.

Float

The purpose of the float in combination with the fuel inlet needle and seat is to control the level of fuel within the Quadrajet's main casting. All other systems of the Quadrajet rely on sufficient fuel quantity in the main fuel bowl to

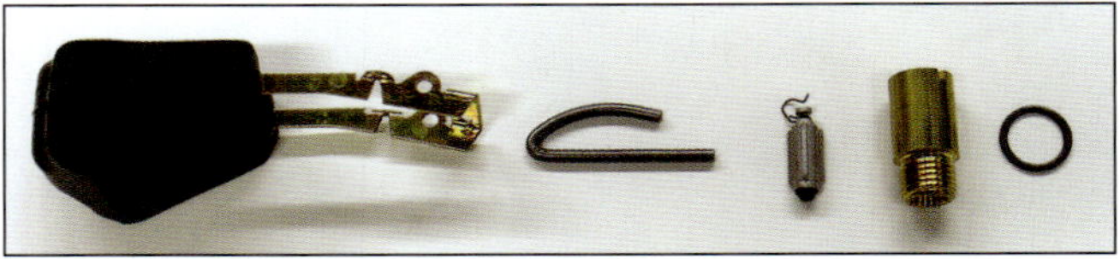

These are the basic components for fuel control for the Quadrajet: (left to right) float, hinge pin (fulcrum), needle, seat, gasket. Although the same basic components were used, the float-and-hinge-pin design would undergo several changes to improve leverage on the needle/seat assembly.

operate correctly during various conditions. The fuel within the bowl is nothing more than a pool which the various systems can draw from. It is essential that close control of the fuel level be maintained. The float itself is nothing more than a pontoon attached to lever arms pivoting on a fulcrum.

The opposite end from the pontoon contacts the needle valve, shutting off fuel flow when the desired fuel level is reached. Quadrajet carburetors used several designs and styles of both floats and needle/seats. Early designs used a plunger/valve arrangement. These were of poor design and quickly upgraded by the factory.

Carburetor number 17059202 (left) is a late style side inlet Chevrolet unit showing the side fuel inlet used on almost all Chevrolet quadrajets. The carburetor on the right has a front inlet and sports carburetor number 17057274, indicating it is a 1977 Pontiac unit.

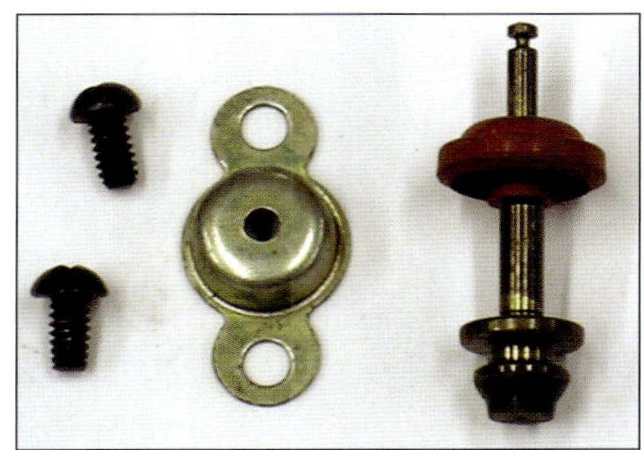

Very early production Q-jets used this type of fuel-inlet design. It proved to be very troublesome and was quickly revised. Rebuild kits are available to update these early carburetors to the later float/fulcrum needle-seat arrangement, and it is highly recommended!

At some point in 1967 these were eliminated due to warranty problems. The castings were redesigned and a large float equipped with a fulcrum attached to the front portion of the power piston housing.

This design was improved again in 1969 by moving the pivot point even further forward. This allowed for an even smaller float to be used.

This turns out to be good news for the Quadrajet, as the fuel bowl is rather small in capacity in the first place. It also allows for more fuel pressure and larger

The 1967 and 1968 Quadrajets all used this float fulcrum design. It was updated in 1969 although the early design would still be produced and used by Oldsmobile and several other applications.

fuel inlet seats to operate effectively in high-performance applications. Even though the basic float fulcrum design was upgraded in 1969, the early design continued to be used in several applications, including Oldsmobile to 1974, many marine applications, as well as various service replacement carburetors.

The Quadrajet float and fuel inlet needle/seat design is extremely reliable and very accurate. Fuel level is critical to correct carburetor operation. Either high or low fuel levels affect the entire calibration of the carburetor, because it becomes difficult to meter fuel when the level is low and easy when the level is high. Typically, low fuel levels cause poor performance and power loss. High fuel levels cause excessive main metering delivery, poor economy and often fuel spillage during vehicle maneuvering

The factory determined the float level for each carburetor. The entire calibration of the carb is based on the fuel level. You may notice that the float level often varies dramatically, even on carbs of the same year and used in similar applications. For stock rebuilding, it is

This 1969 carburetor shows how the float fulcrum was moved forward in the casting with a smaller float employed. This was a good move for the Quadrajet and improved fuel control while freeing up more room for fuel in the main bowl.

The float fulcrum was redesigned again for the 1975 and newer carburetors. This allowed for an even smaller float to be used while still maintaining sufficient pressure on the needle/seat assembly.

Setting the float level is critical to correct carburetor function. The factory settings should be followed exactly for stock rebuilds.

The idle-mixture holes beneath the mixture screws exit directly under the primary throttle plates. The vertical cuts in the casting just above the idle mixture screw holes are the transfer slots. They supply some idle fuel as well as just off-idle fuel as the angle of the throttle plates is increased.

always best to follow the manufacturer's recommendations for float level, since float settings, fuel inlet seat diameters and fuel pressure have a direct impact on the level of fuel in the main casting.

Idle System

The purpose of the idle system is to supply the engine with the correct Air/Fuel (A/F) mixtures at curb idle and low-throttle openings/vehicle speeds. In this operating range, air flowing through the primary bores is insufficient in velocity to create the necessary pressure drop to pull the needed fuel from the main nozzles. Therefore the idle system must supply fuel under these conditions.

The idle system is activated when the primary throttle plates are nearly closed. This causes a high-vacuum situation beneath the primary throttle plates. The idle mixture holes under the mixture screws exit beneath the plates. The vacuum present at idle and low speeds causes fuel to flow from the holes as internal passages in the base plate and main body of the carburetor connect with the fuel present in the bowl.

The idle system is somewhat complex, and various designs were used for

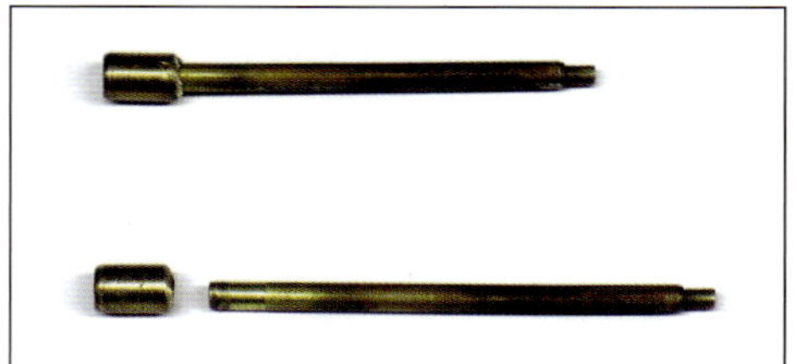

The idle tubes are actually two separate parts containing a tube with a restriction on the small end and a collar pressed onto the other end. The idle tubes are the first restriction in the idle system and hang into the main body well below the fuel level. They should always be removed from the main body for any level of rebuilding since dirt and debris is typically found in and under them.

different models and applications. All Quadrajets use idle tubes pressed into the main body.

They are located just inboard and forward of the two screw holes inside the choke housing on each side of the primary bores.

The brass tubes are actually two pieces consisting of a tube and a collar. At the bottom of each tube is a restriction. The size of the restriction is critical to correct idle fuel delivery. The size of the

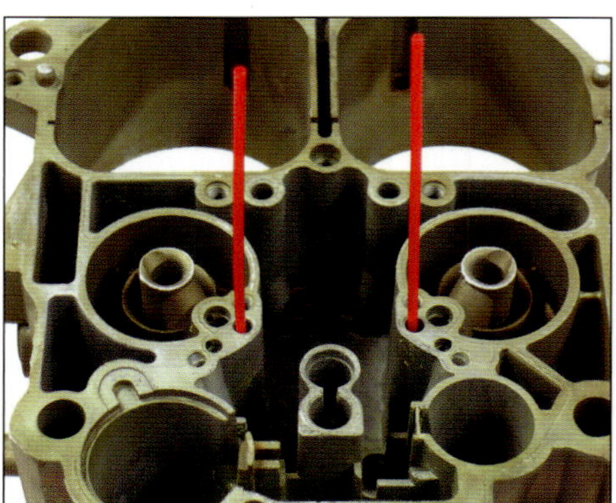

The idle tubes are located on the primary side of the carburetor just forward and outboard of the two screw holes found inside the choke housing. They share a common passage with the idle-down channel holes, located just forward and outboard of them on the main body.

restriction in the idle tube varies greatly for different applications. The other parts of the idle system that we discuss are also sized accordingly. This makes each factory Quadrajet very application-specific and is the single biggest reason why they often get a bad reputation as a high-performance carburetor. The restrictions in the idle system on factory units almost always fall short of supplying sufficient idle fuel and bypass air to heavily

The air-horn gasket has a cutout between the idle tubes and down-channel locations in the main body. This allows fuel that is pulled up through the idle tubes to turn and be pulled down to the base plate and idle-mixture screws.

This fact is very critical as it makes swapping parts around between different carburetors difficult, unless both original units used the same type and size of idle air bleeds. The idle air bleeds allow air to be mixed with the idle fuel for improved atomization. They also have a great effect on the signal to the idle tubes and fuel delivery. Some carburetor models that used idle air bleeds in the air horn also have an adjustment screw beneath a triangular metal cover located in the front of the choke housing either over the front vent, or on the boss where it would normally be for external vented units.

There is an adjustment screw beneath the cover that allows air to enter

cammed engines, and in many cases are insufficient if used on any engine much outside of the original application.

As fuel is pulled up through the idle tubes and exits, it is subjected to an idle air bleed. Some Quadrajets have the idle air bleeds located in the air horn.

Other models have them located in the main casting just below the gasket surface and within the adjacent passage that leads to the idle channel restriction.

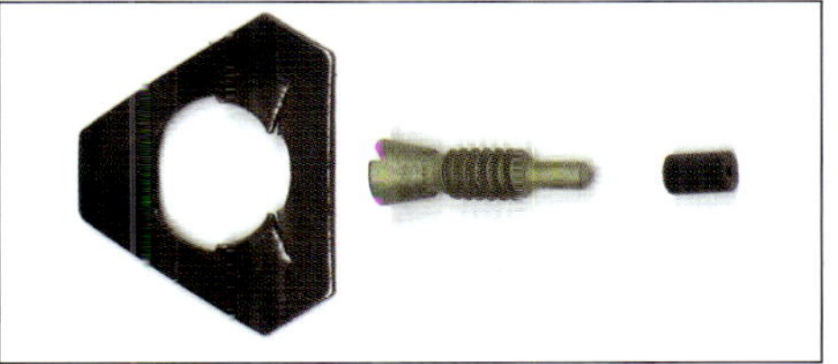

Some air horns have a triangular-shaped metal cover over the front vent or where the front vent would have been on some models. Under the metal plate is an adjustment screw. Two passages lead from under the screw to just below the idle air bleeds, making them adjustable. (On some 1970 to 1974 carburetors the passages may lead under the main air bleeds instead.) The top photo shows the adjustment screw, cover and tiny rubber sleeve used to retain the adjustment screw. The photo above shows a close-up of the adjustment screw in the air horn with the vent and cover removed.

Some models use idle air bleeds in the air horn as shown in this 1969 Pontiac casting. The idle air bleeds expose the idle fuel that has been pulled up through the idle tubes to incoming air before it proceeds down through the idle channel restriction and makes it to the base plate.

Some models use idle air bleeds in the main casting as seen on this late 1970's Chevrolet casting. The carburetor can only use one pair of upper idle air bleeds, either in the air horn or main body, not both. This puts limitations on parts swapping without modifying one part or the other. A small drill bit has bit inserted into the idle air bleed to show the passage of air from the primary bores to just above the idle down channel restriction.

beneath the idle air-bleed holes in the main casting. These particular carburetors provide additional flexibility for tuning the idle system externally.

As the idle fuel passes the idle air-bleeds, it is pulled downward through a passage leading to the base plate. In this passage, just below the gasket surface, we find the idle channel restriction. It is a pressed-in brass cup with a sized hole in the center.

The size of the idle channel restriction plays another important role in the

The lower idle air bleeds supply some additional air to the idle fuel as it travels down through the main body on its way to the idle-mixture screws. Lower idle air bleeds also serve an additional function. They supply some additional fuel at increased throttle openings through wide-open throttle to supplement the main discharge nozzle delivery.

Fuel passing up from the idle tubes is pulled over and down through a passage leading to the base plate. The idle channel restrictions are pressed in brass cups with specific-sized holes to control the amount of fuel metered to the rest of the idle system. Idle air bleeds are located just above the idle channel restrictions if not located in the air horn. The carburetor uses one type of upper-idle air bleed only, not both.

Located in the base plate are two drilled passages that exit just beneath the front center portion of the primary bores, which are just below the throttle plates when they are in a closed position. Mixture screws intersect these passages and are adjustable externally as the final control of how much fuel actually enters the engine at idle/low engine speeds. Directly in line with the idle-mixture screws are off-idle discharge ports or slots. These slots extend both above and below the throttle plates when they are nearly closed. In part they supply some idle fuel, and as the throttle blade angle is increased, some off-idle fuel. As the primary throttle valves are opened just off idle to increase engine speed, the off-idle discharge ports are subjected to high vacuum. This provides additional fuel to the extra air entering the engine. The off-idle discharge holes or slots help with

amount of fuel allowed to the mixture screws during various engine vacuum situations. It is another very important tuning area for high-performance applications.

Also located in the main body and intersecting the idle down channel passage are the lower idle air-bleeds. Much like the upper idle air-bleeds, these holes supply some air to the fuel, but it is mixed after the idle down channel restrictions.

Once past the lower idle air-bleeds, fuel finds its way to the base plate.

Here's a close-up of a base plate with the throttle plates removed showing the idle discharge holes and the off-idle discharge ports. The idle discharge ports are actually slots cut into the throttle plate. The position of the primary throttle plates at idle speed has the idle discharge holes and a small portion of the slots exposed. The high vacuum present at idle pulls fuel from both locations. The idle-mixture screws control the amount of fuel flow through the idle discharge holes. Since fuel continues to flow uncontrolled from the exposed portion of the discharge ports, it may not always be possible to kill off the engine by seating the idle-mixture screws. Even so, on a correctly calibrated carburetor, seating the idle-mixture screws should slow the engine considerably.

the transition from idle to the main system and are a very important part of the Quadrajet function.

Another part of the Quadrajet's idle system is idle bypass air. Not all Quadrajets use bypass air, and not all of them have the system completely intact. The components and passages are present on most units, and it can be employed if needed. On those units that don't use it or don't have all of the passages present, the labor and expertise involved to employ it might be better spent on locating a unit with the system intact. Even so, the vast majority of Quadrajets produced through the years of production have bypass air or all of the passages in place.

Bypass air is nothing more than a precisely controlled vacuum leak. It is important to the Quadrajet, as it allows the throttle plates to be closed further at idle when needed for certain applications. The small and efficient primaries used by the Quadrajet cause fuel flow from the main nozzles at relatively low throttle angle and engine vacuum. When engine vacuum is low at idle (typically caused by installing a cam with a lot of overlap and/or a tight lobe separation angle), it is often necessary to increase the angle of the throttle plates, allowing enough air to enter the engine to sustain it. With the Quadrajet in particular this can cause fuel to start flowing from the main nozzles. This is often called the famous Quadrajet "nozzle drip." Nozzle drip also occurs simply because the idle system is too restrictive for the application and the engine needs more fuel to just keep it running. Even so, the idle bypass

air system is instrumental in controlling primary throttle angle at idle speed even when the amount of idle fuel is sufficient. Typically, nearly "stock" engines with small cams and lots of vacuum at curb idle don't need or want idle bypass air. From "mild" applications right up to full "race" engines, idle bypass air in conjunction with the rest of the idle system being well tuned becomes important to getting a strong, stable idle in and out of gear.

Idle bypass air is very similar in function to idle fuel except it simply reroutes air only. The air passes through the carburetor then below the throttle plates at idle. The air used for idle bypass air enters through the primary side and is

Idle bypass air was used on some models through the years of production. These photos show two different styles of base plates and the location of the idle bypass air holes under the throttle plates.

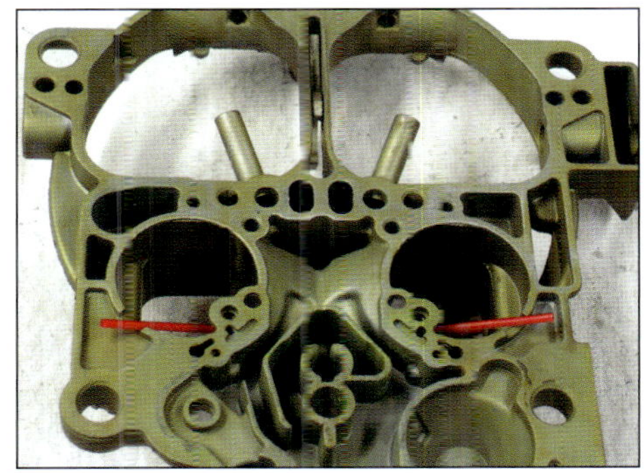

The slot in the air horn allows filtered incoming air to enter voids on either side of the main fuel bowl.

The bypass air enters the voids and continues on to the base plate to the bypass air holes. The correct gasket must be used to line up the holes in the base plate to the voids. This is because several different styles of base plates were used, even during the same model year.

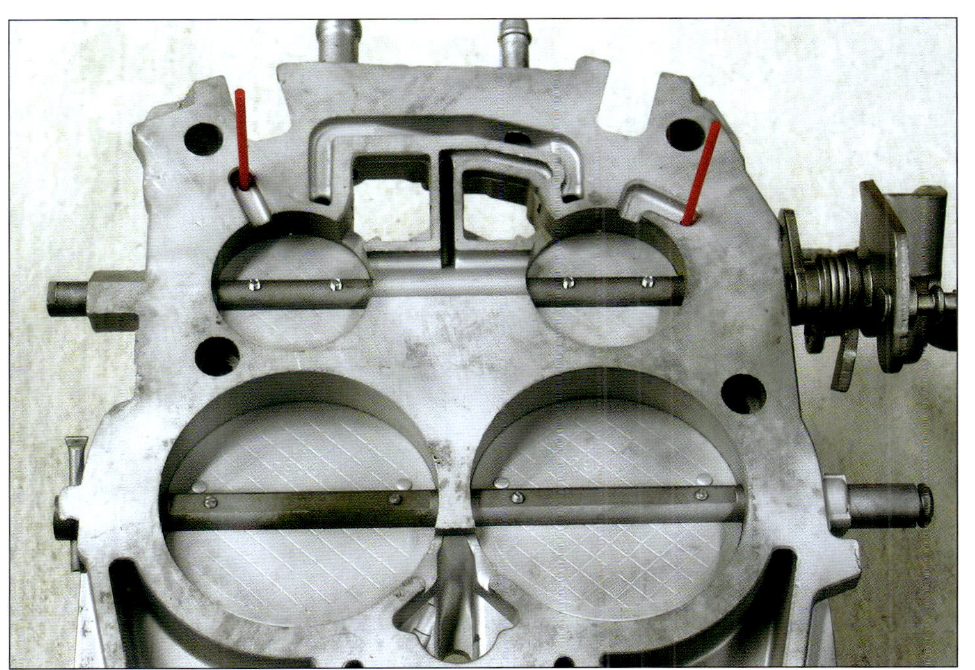

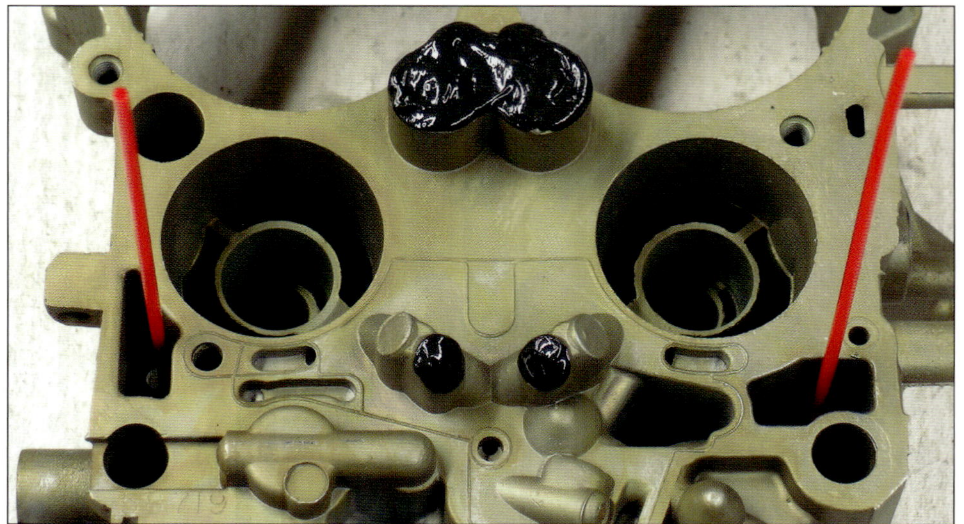

This bottom view shows the holes in the bottom of the voids that lead to the corresponding holes in the base plate.

Idle bypass air that comes down from the voids enters the base plate and then moves on to the passages exiting under the throttle plates.

In addition to idle bypass air, some models also had drilled throttle plates to further reduce the required angle of the plates and maintain idle speed.

pulled in through two slots in the air horn. It then goes over and down through two voids right and left of the main fuel bowl. Passages in the base plate, if present, exit just beneath the throttle plates. Some base plates use an angle-drilled passage that intersects a vertical drilled passage; others use channels as the exit holes are drilled directly in the bottom of the vertical passages. Either system works equally well.

Some factory "high-performance" Quadrajets not only used idle bypass air but additional ³⁄₃₂-inch holes drilled directly in the front of the primary throttle plates in line with the idle mixture screws. These holes also helped to reduce throttle angle at idle.

The idle system and its components are the most important part of the Quadrajet in terms of high-performance modifications. Study the terms and photos closely

and become very familiar with them. In the high-performance section (Chapter 6) a large part of the effort to set up your carburetor is spent on the idle system.

Primary Main Metering System

The Quadrajet carburetor is classified as a spread bore design. Being of the spread bore design, the primary side of the Quadrajet uses small bores. There were only two sizes produced through the years of production. All Quadrajets prior to 1971 use the smaller bores. In 1971 larger primary bore castings started showing up. In addition, Pontiac introduced a Quadrajet lacking the outer primary booster rings. Both were attempts to get more CFM from the original design. The Pontiac carburetors were limited in use, and due to their lack of efficiency at very light throttle openings were dropped by 1972, being very difficult to tune for emissions. In 1973 and 1974 Pontiac used the larger primary bore castings exclusively on the rare and highly sought-after Super Duty carburetors. Buick continued to use the large primary bore castings through 1974; these are the only divorced-choke large CFM castings produced. In and after 1975, all front inlet non-ECM Quadrajets use the larger castings. The vast

The larger main primary-bore carburetors, shown on the left, started showing up in the early 1970s. They can be easily identified by the "bump" in the casting noted at the 10 o'clock position in this photo. The smaller casting shown on the right also has a much more pronounced "ridge" and noticeably smaller opening.

Here are two early samples of the larger main-body carburetors. On the left is the rare and valuable 1973 Pontiac Super Duty carburetor. On the right is a Buick 704X240 carb. Note that the Super Duty carb uses the hot-air-style choke (choke housing attached to the carburetor) and the Buick model has a divorced choke. Both models also have slashed front vents although they are different styles.

A triple-venturi area is formed by adding a booster ring around the main fuel nozzle (center of picture). The size and shape formed by the three distinct areas provide an excellent signal to the main fuel system to help with precise fuel metering.

lent fuel atomization. The triple venturi principle also provides flow from the main nozzles at very low throttle openings. This provides a seamless transition from the idle to main primary fuel system for "normal" driving.

The point at which fuel starts flowing from the main nozzles is called the "transfer point." At the transfer point both the idle and primary main system work together to supply the necessary air/fuel ratio. The primary main system continues to supply fuel past the transfer point up to and including wide-open throttle operation. All of the fuel that is delivered to the main nozzles flows past a main metering jet.

majority of the side inlet castings are the larger bores. Even so, some smaller displacement engines would receive the smaller primary bore carburetors in and after 1975.

Regardless of whether the large or small primary bores were used, the approach to tuning the primary side of the carburetor is the same. The small primaries allow high-velocity air to cross the venturi area of the carburetor at relatively low engine speeds and airflow. The velocity of the air in conjunction with the excellent nozzle and booster arrange-ment make the Quadrajet unsurpassed in efficiency. With the exception of the 1971 Pontiac "HO" Quadrajets, the primary booster is surrounded by a cast circular ring. The booster "ring" creates a third venturi area around the main fuel nozzle. A triple venturi area is created between the main body, booster ring and main nozzle. Considering the relatively small size of the primary bore in the first place, having the additional venturi areas in place puts a lot of "pull" on the main fuel nozzles. This makes for very precise metering from the main jets and excel-

The primary main jets are well located in the bottom of the main fuel bowl as shown.

A metering rod is used to vary fuel delivery to the primary main fuel system. Notice that the metering rod has two different sections and the upper section is tapered. The position of the metering rod in the jet varies the air/fuel ratio for various engine-operating conditions.

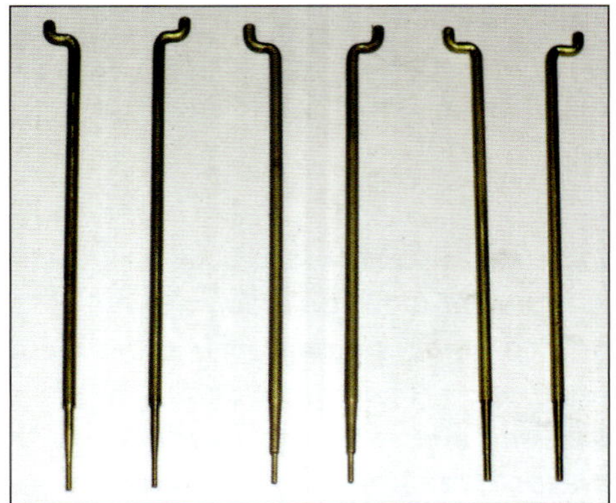

There were quite a few different metering rods used through the years of production. Many have a tapered second section and the height in the jets may have been adjustable for very fine metering control.

sections to control the amount of fuel that passes by the jets at both high- and low-vacuum situations. Some models use two step-rods with a tapered section between steps; others use a tapered second section. There are also two different tip sizes used for primary metering rods, .026 inch and .036 inch, and two different lengths depending on year and application.

The size of the jet in conjunction with the size of the main air bleed(s) control the amount of fuel delivered to the nozzles. There is either one or two pair of main air bleeds. Most early units use two pairs of main air bleeds, one pair located in the air horn directly over the primary main fuel supply passage and a second pair in the main body intersecting the main fuel passages at a 45-degree angle.

In the 1970s a second design of Quadrajet was developed with a modified main body and air horn that use only one main air bleed. This model of carburetor was produced before and after 1975, and requires a different gasket that includes a "teardrop" shape at the main nozzle fuel passage. The main air bleeds on these models are located in the air horn only. They are brass tubes pressed into the air horn that extend

A wide variety of different main air bleed sizes were used. The picture on the left shows the main air bleeds in the air horn. The picture on the right shows the location of the main body air bleeds. Some models have main air bleeds ONLY in the air horn (top photo); others may have one pair each in the air horn and main body. The size of the main air bleed(s) directly affects final metering, and the jets and rods were sized accordingly.

upward at a slight angle protruding outward into the main air stream.

The Quadrajet also uses metering rods with two distinctly different-sized

A power piston controls the height of the metering rods in the jets. The power piston rides within close tolerance to a bore within the primary side of the

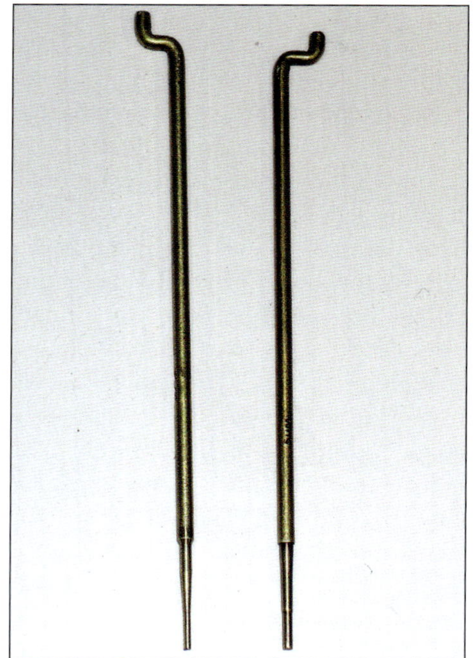

In addition to using a wide variety of primary metering rods that may also feature tapered second sections, some use .026-inch tips and others .036-inch tips. The larger tip rods are more common in the late 1970s through the 1980s.

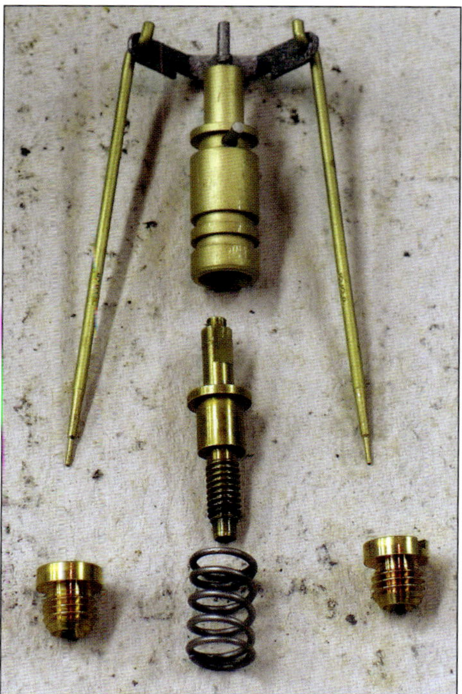

The power piston sits in a bore in the center of the primary side of the carb. A small hole at the bottom of the bore leads to a corresponding hole in the base plate. It is exposed to manifold vacuum at all times. A spring raises the power piston whenever engine vacuum falls below the rating of the spring. This raises the power piston and also the metering rods within the jets to the smaller end section, increasing fuel delivery. Some Quadrajets that were delivered on engines equipped with turbochargers do not have the passages present in the base to always expose the power piston to manifold vacuum. On these applications, manifold vacuum is routed externally so that when the engine goes under boost from the turbocharger it can be shut off to keep from pressurizing the fuel bowl.

carburetor that is subjected to engine vacuum below the throttle plates, or manifold vacuum. There is a spring beneath the power piston that causes the piston to rise during low vacuum situations. This typically occurs at heavy/full throttle, and the upward movement of the piston positions the smaller or bottom section of the metering rod within the jet for increased fuel delivery. During high-vacuum situations found at idle and light part-throttle cruising, the vacuum is sufficient to overcome the spring tension and the power piston is lowered. This positions the second or larger portion of the metering rod within the main jet and decreases fuel delivery.

There were several different power piston designs used through the years of production. Some employed the use of an adjustable part throttle (APT). Three different APT designs are discussed in greater detail in Chapter 6. Quadrajets that have APT give the user full control of the part-throttle A/F ratios externally. This is a very desirable feature.

The Quadrajet also uses an accelerator pump to add additional fuel to the primary side of the carburetor when the throttle is moved quickly. Any quick movement of the throttle immediately allows additional air to enter the engine. The engine vacuum also falls off. The combination of additional intake air and low vacuum causes a momentary lean condition as it takes some time for the fuel to begin flowing from the main nozzles. To offset the lean condition, a pump is attached to the throttle linkage. Any quick throttle movement depresses the pump and plunger and attempts to compress all of the fuel under the pump seal. The fuel quickly discharges into the primary bores through two small discharge holes in the air horn. A checkball is used in the bottom of the main casting to keep fuel present all the way to the discharge holes in the air horn. The size of the holes varies slightly depending on carburetor model and application. The size of the holes controls the amount and duration of the fuel that is forced into the primary main bores during quick throttle movements. The accelerator pump is a key player in setting up a high-performance carburetor and must deliver the correct amount of fuel so the engine doesn't stumble, hesitate, or falter when one goes quickly to full throttle.

Some Quadrajets, both early and late styles, use an additional fuel-enrichment system to assist the accelerator pump in delivering a quick shot of fuel to the primary side of the carburetor.

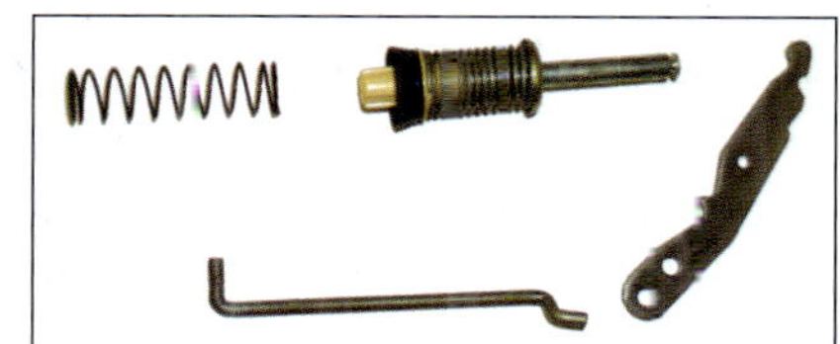

The basic components of the accelerator pump are shown here. The spring under the pump keeps pump in constant contact with the lever that applies it. The spring on the pump allows it to compress the fuel trapped under it without breaking the pump or blowing out the pump seal.

<hr>

Fuel trapped beneath the accelerator pump is forced through internal passages in the carburetor to the air horn, where it passes through very small holes into the primary main bores adjacent to the primary boosters. A checkball in the bottom of the carburetor is used to keep the carburetor's internal passages full of fuel at all times to avoid a delay in fuel delivery when the pump operates.

The accelerator-pump discharge holes are located in the air horn and vary in size between different models.

Early designs have fuel passages in the air horn with drilled exit holes within the choke housing. These holes are connected to tubes that extend from the bottom of the air horn into the main secondary fuel passage. Later carbs use a separate set of tubes in the air horn that hang in the main fuel chamber on each side of the float. They also have tiny drilled holes exiting into the primary side of the carburetor. With either system additional fuel is supplied to the primary side of the carburetor upon any quick movement of the throttle. The holes are quite small to control the amount of fuel delivered. These systems were employed to assist the accelerator pump and to allow for leaner part-throttle calibrations without lean spots during transition from the idle to the primary main system. Extensive street, dyno, and track testing has shown that this system is not needed and shows no improvement in performance on a well-prepared carburetor.

Secondary System

The secondary side of the Quadrajet is used to provide large amounts of air and fuel on demand for full-throttle situations. Being of the spread-bore design, the secondary throttle plates are huge, measuring 2¼ inches each. The throttle plates themselves are mechanically activated by a link from the primary throttle shaft. When the primary shaft reaches about ⅔ of its full-open position, the secondary throttle plates begin to open and achieve a full-open position at the same time as the primary side. Some models were intentionally adjusted so that the secondary throttle plates do not reach a full-open position. This was more common with later emission years' carburetors typically found on small-displacement engines.

Located in the air horn are two air flaps on a common shaft. They control all of the air that is allowed to enter the carburetor whenever the mechanical

Although not used on all models, both early and late production carburetors were delivered with a primary fuel-enrichment system. Tiny holes in the top of the air horn provide fuel in addition to what is supplied by the accelerator pump during quick operation of the throttle. On the late-style carburetors produced in and after 1975, the air horn has an additional set of long tubes if this system is used.

secondary throttle plates in the base plate open. The air flap shaft is held closed by an adjustable spring. The spring not only keeps the air flaps closed but also controls the opening rate during full-throttle operation. If the huge mechanical secondaries were quickly opened, the resulting incoming air would cause a huge bog, hesitation, stumble, or even backfire up through the carburetor, due to a lean condition. The air flap is used to provide control of the air that enters the carburetor. Its progressive operation ensures smooth engine operation during full-throttle application. The outer portion of the linkage is connected to the choke pull-off, assisting the spring to ensure the flap does not "whip" open too quickly. The choke pull-off holds the secondary air flaps tightly shut during high-vacuum situations and allows the flaps to open when the vacuum falls off at full throttle. The spring tension and choke pull-off opening rate are critical to correct full-throttle performance and are discussed in greater detail in Chapter 5.

All of the fuel flowing to the secondary side of the Quadrajet flows past two pressed-in steel disks in the bottom rear portion of the fuel bowl. The fuel is drawn up through two passages on each side of the main body. Two long brass tubes hang into each fuel passage and provide air to help fuel atomization before it reaches the main nozzles.

All Quadrajets used mechanical linkage to operate the large secondary throttle plates. There are several different versions, but all work in the same manner.

Even though the secondary throttle plates are mechanically operated, large air flaps in the air horn control how much air enters the engine. These air flaps ride on a shaft kept tightly closed by spring pressure and a link from the choke pull-off.

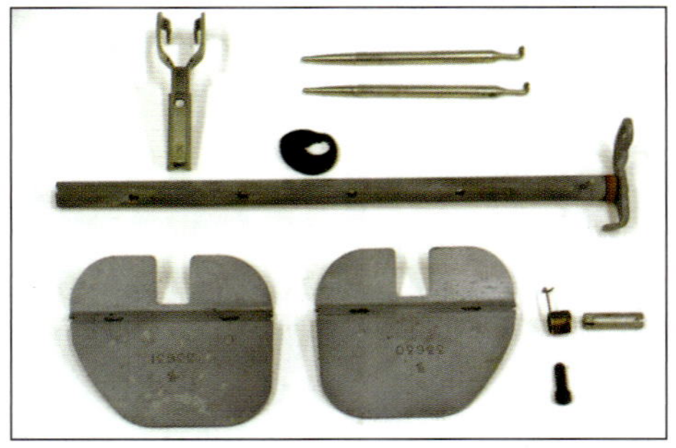

The secondary air-flap shaft and related parts are shown here. The plastic cam raises the secondary metering-rod hanger. Tapered secondary metering rods are used. Through its action, the cam increases fuel delivery proportionally to the increase in incoming air as the engine's speed increases.

An additional enrichment system provides additional fuel at the moment the large secondary throttle plates open. This acts much like the accelerator pump does for the primary side of the carburetor, preventing a lean condition before fuel is flowing well from the secondary main nozzles. Some very early Quadrajets do not use this system, but with some effort it can be added.

Two fixed-orifice disks are pressed into the main fuel bowl. All fuel that enters the secondary side of the carburetor flows past these disks.

The main nozzles are ¼ inch inside diameter and extend at an angle down into the center of each secondary bore. A divider plate is present and the main fuel nozzles extend just pass the divider. Secondary metering rods control the amount of fuel delivered to the main fuel nozzles. The rods are attached to a hanger that is raised via a small plastic cam in the center of the secondary air flap shaft. When the air flaps are closed the rods hang the furthest into the pressed-in disks in the bottom of the main fuel bowl.

The rods are tapered so that any upward movement of the hanger raises the rods to a smaller portion and allows more fuel to pass by the disks. The progressive operation of the spring-loaded air flap and the use of the cam, hanger, and tapered rods ensure the correct fuel delivery as the engine RPM increases at full throttle. Correctly calibrated, this system provides seamless full-throttle performance and flawless transition throughout the load and RPM range of the engine. All of the components, the hanger, rods, and choke pull-off, and air

flap tension spring are accessible without taking the carburetor apart. This makes tuning the secondary side of the Quadrajet extremely easy and takes only a few minutes.

Except on some very early 1967 and 1968 carburetors, an enrichment system is present to assist the secondary side of the carburetor with fuel delivery. The system acts much like the accelerator pump on the primary side of the carb, but is activated by the quick pressure drop created by the opening of the huge secondary throttle plates. The enrichment, or the "pull-over" system is designed to eliminate a lean condition that would be present as the secondary throttle plates open and vacuum is applied to the air flap. The fuel to the pullover system comes from two voids on either side of the rear portion of the main fuel bowl. Tiny holes lead from the main fuel bowl into each void. The holes are very small in size, typically between .035 and .040 inch.

They are designed to control fuel delivery when the secondaries are first opened, then limit fuel delivery after sufficient fuel begins to flow from the main nozzles. Two tubes extend down from the air horn into each void and lead into two holes that exit into the secondary

to supply a small amount of fuel to avoid a lean condition at the onset of secondary operation. Tuning this system is critical to smooth full-throttle performance and is one of the most important features of the secondary system.

Through the years of production the factory would limit the opening angle of the air flap to control the total CFM of airflow through the carburetor. Many late small-cubic-inch applications have

the flap-opening angle restricted so much that additional holes were used in the main fuel nozzles to ensure sufficient fuel delivery.

Choke System

The choke system is used for easy cold-engine starts and to provide the richer mixture required during engine warm up. Fuel condenses quickly on the

Two fuel wells are present on each side of the main fuel bowl in front of the large secondary openings. Two tiny holes allow fuel to enter and fill the wells to the same level as fuel in the main bowl. A brass tube from the air horn hangs into each well, pulls fuel, and discharges it into the secondaries any time they are quickly opened. On some very early models the voids are present but may not be drilled because the rest of the enrichment system was not used.

The secondary enrichment and all of its components are part of the air horn. Shown here is a Pontiac air horn that used the notched flaps to expose the fuel-supply holes. Most air horns have smaller holes located above the secondary air flaps.

side of the carb. Most of these holes are located above the air flaps. Some Pontiac carburetors have notched flaps to expose these holes to the underside of the air flaps. A few other models have the holes located under the air flaps. Regardless of the location, the purpose is

At the end of the secondary air-flap shaft, the linkage has a stop to limit or control the maximum open angle. The factory used this to control the CFM of the carburetor closely for specific applications.

The full open point of the secondary air flaps was so limited on many late-model carburetors that additional holes were added to the fuel-supply tubes, ensuring adequate fuel delivery at full throttle.

intake manifold runners when the engine is cold. In order to ensure that a combustible mixture reaches the cylinders, the Quadrajet is equipped with a choke flap located in the air horn just above the primary throttle bores. When the choke flap is closed during engine cranking, very little air is allowed to enter the engine, and the engine vacuum created is applied directly to the

The choke flap is mounted on a shaft with linkage connected to the choke assembly on the right side of the carburetor.

idle and main metering systems. Although this is desirable for initial start up, the air/fuel ratio must be leaned up somewhat once the engine starts. A choke pull-off is used to unload the choke to a specified angle after initial start-up.

The choke flap is controlled by a thermostatic coil that is affected by temperature differences. Cold temperatures cause the coil to wind in a direction that applies the choke, and it retracts and opens the choke as the metal coil is warmed. Quadrajets basically have two different choke designs. The choke coil is either mounted on the intake manifold (divorced), or in a housing attached to the carburetor. The divorced choke coil sits within a small metal cover directly over the intake manifold heat crossover passage. A metal link attaches the choke coil to the linkage on the carb that controls the angle of the choke flap and the position of the fast idle cam. Hot exhaust gases present in the crossover passage quickly warm up the divorced choke coil to unload the choke as the

engine warms up. If mounted on the carburetor, the choke is activated by hot air supplied through a tube that routes air into the heat crossover passage in the intake. The intake air for the hot air style chokes is routed through the rear portion

Most of the later production carburetors used an electric choke instead of divorced or hot air. This provided additional flexibility and easily lets any Quadrajet utilize a choke without using the engine as a heat source for correct operation.

There were two basic designs through the years of production. The carburetor on the right uses hot air from the intake manifold cross-over passage to heat the choke coil. The carburetor on the left contains the choke linkage and fast-idle cam, but the choke coil is mounted on the engine. These units are typically referred to as using a "divorced" choke.

of the air horn just behind the secondary hanger. Using filtered fresh air ensures that the hot air supplied to the choke is not contaminated with carbon and other elements present in the exhaust gases that would clog up the choke and make it inoperable. Starting in the late '70s, the hot-air-style choke coils were replaced on some models with an electrically activated coil. The electric coils can be installed on any year of Quadrajet using the hot-air-style coil, provided the vacuum to the housing is blocked off and the gasket removed as the unit uses the housing as a ground.

Regardless of the type of choke used, the basic operation is pretty much the same. When the engine is cold, the operator must depress the accelerator pedal to engage the choke. The choke coil provides sufficient pressure to close the choke flap and at the same time raise the fast idle cam to the highest position. Depressing the accelerator also delivers a shot of fuel to the engine via the accelerator pump in the carburetor. It is not uncommon to depress the accelerator two or three times on cold starts to supply the engine with enough fuel for quick starting. Once the engine starts, it quickly pulls enough fuel from the fuel bowl, and no further engine priming is required.

The choke pull-off unloads the choke to a specified angle, depending on carburetor model. Later units used a secondary pull-off to assist the primary pull-off, ensuring the choke flap angle was increased sufficiently to avoid loading up the engine with an overly rich air/fuel mixture.

As the choke flap opens during engine warming, the fast idle cam falls to lower steps to decrease engine RPM. There were many different setups used during the years of production, yet all of them work pretty much in the same manner.

The choke can be removed for high-performance applications. This is a common practice since many race engines are only operated in warmer weather and are typically set up with more generous idle and off-idle fuel curves. In addition, a large majority of muscle cars and street/strip vehicles are not operated in colder weather. This greatly diminishes the need for a choke in the first place. The closer the carburetor is tuned for the application, the easier the engine starts and continues to run without the need for an overly rich mixture. It is also a common practice to either block off or fill the exhaust crossovers in the cylinder heads for a denser and richer fuel mixture for increased power. Many aftermarket cylinder heads do not even have an exhaust

Many later production carburetors also used a secondary choke pull-off to further assist in choke unloading after the engine was started.

Some models released in the late 1970s through the early 1980s used the rear choke pull-off only and two linkages for both choke unloading and to hold the secondary air flap shaft closed at idle and cruising.

crossover in the first place. Both these conditions require either a hand choke or an electric choke conversion if the vehicle owner desires correct choke operation for cold start-ups.

A choke has not been used for many years on my own vehicle, and it is daily driven except in extremely cold conditions. The choke is not even missed, as the engine still starts easily and idles fine within a few minutes of operation. At the time of this writing at least 30 percent of the carburetors that we are building do not include a choke. About the same percentage have the later carbs with the housing on the carburetor converted to electric, and most of those do not have the choke flap and shaft in place. This provides fast idle capability for convenience without loading up the engine with an overly rich mixture during starting and engine warm-up.

Here is a Quadrajet with the choke flap and shaft removed. In addition to not having any choke operation for cold starting, removing the components offers some additional airflow often needed for high-performance applications.

SELECTING THE RIGHT CARB

In previous chapters we have covered the history and evolution of the Quadrajet and the basic systems and how they work. Before moving on into rebuilding and high-performance modifications, you need to make sure that the carburetor you have chosen meets your basic requirements so your finished product works and performs as expected.

The good news is that the basic design of the Quadrajet stayed the same throughout the years of production. The Quadrajet would receive numerous minor changes. Some of these changes were a direct result of ever-tightening emission standards. Others were to make the basic design more flexible and reliable. Most of the changes were for the better, and as you will see, the later production units sport a lot of very good features, making them an excellent starting point for a high-performance carburetor.

The use of the Quadrajet through its years of production was so broad that countless used samples are currently available and will be for many, many years. It has long been thought that the only carburetors worth working with were the rare and highly coveted factory high-performance carburetors. This has had countless enthusiasts searching the salvage yards and

swap meets for years, drying up just about all of these units. The good news is that the factory "high-performance" carburetors were only slightly reworked standard production units. This means that, armed with the right knowledge and abilities, we can rework our standard production Quadrajet to virtually any level of performance needed.

The carburetor you choose to work with still has to meet some basic requirements for the application. There are several minor things to consider, such as fuel inlet location, CFM capability, linkage, vacuum fittings, etc. We also have to look at the year of production and overall condition of the carburetor. In many cases it is more cost-effective to

These two 1969 Pontiac carburetors look very similar. The carburetor on the right is a factory high-performance carburetor used on 400ci Ram Air engines. The carburetor on the left was used on the standard 400 engine the same year. Both have slashed front vent tubes. The "270" Ram Air carburetor used a slightly different calibration than the standard production "262" carburetor; otherwise they are nearly identical.

choose a later production unit in better condition than to spend a lot of hours working with a well-worn early model.

What Are the "Best" Quadrajets?

I have been asked this question countless times. The most accurate and direct answer would be the factory high-performance carburetors. The very best factory Quadrajets would be the 1973 and 1974 Pontiac 455 Super Duty carburetors.

In 1973 and 1974 Pontiac used a special casting with larger primary bores for their Super Duty 455 engines. These carburetors have excellent idle-fuel calibrations along with very generous fuel curves. They can be quickly identified by the large slashed front vent tube. They are excellent high-performance units, but extremely rare and highly coveted by restorers.

They were among the first to use the larger castings. They also had an excellent secondary fuel-enrichment system and generous idle and off-idle fuel settings. These particular carburetors also used the hot-air-style chokes and can easily be converted to electric choke, making for easy retrofitting on most any application. They work quite well in near stock form on most any high-performance application. The problem with them is that they were produced in very limited quantities; the Pontiac Super Duty Trans Am was quite rare. Even ser-

vice replacement Super Duty carburetors are rare and bring high prices.

Next in line would be the 1971–1974 Buick carburetors delivered on the 455 engines. These carburetors also used the larger castings, yet were divorced-choke. They make excellent carburetors for use on any early factory

intake that had a divorced choke set-up, particularly where the user has swapped in a high-performance or large-CID engine. The large-CFM Buick carburetors are relatively common and can be purchased at reasonable prices.

Following closely behind the large-CFM Pontiac Super Duty and 455

Many 1971–1974 Buick carburetors also featured the larger castings. These were the only divorced-choke, large-CFM castings produced. They can be easily identified by the slashed front vent tube and large "hook" on the throttle linkage. They make excellent carburetors for any divorced-choke application; however, they lack any provision on the throttle linkage for transmission cables and return springs behind the carburetor.

In 1971 Pontiac used special castings lacking the outer booster rings. These were early attempts to increase CFM capabilities for their 455 HO engines that used the new round-port heads. They only survived one year. Lacking the outer booster rings decreased the signal to the main jets. The lost efficiency made them very difficult to get past the emissions standards in place at the time.

1975 APT Carburetors

New castings showed up in 1975 featuring an early APT system. Located in the right front corner of the carburetor is a brass cup plug covering an adjustment screw. Turning the adjustment screw provides fine control of the part-throttle mixtures externally.

Some 1975 and 1976 Quadrajets also used a second power piston and single jet/rod in the main body. Although these carburetors have the same potential as any other late-style Quadrajet, they are more labor intensive to recalibrate correctly. Due to the increased effort, it may be a better idea to simply get another carburetor that doesn't use this system. The single APT carburetors have internal passages that route additional fuel beneath the main jets. The factory pressed a single jet into the main body of the carburetor and added a single metering rod for fine external metering control. The problem with this particular design is that raising the metering rod not only adds fuel at part throttle, but does so at heavy and full throttle. This would have the engineers compensating for the increased fuel with main system calibration. I'm sure this is not exactly what they intended, and it met with mixed results as the design was only used for a brief period, then upgraded with the later APT that raised the power piston only.

The later design raised only the power piston and used metering rods with a tapered second section. This provided very fine control of the part throttle air/fuel ratios without affecting heavy and full throttle metering. This later design was used throughout the remaining years of production for non-computer managed carburetors.

In 1976 the APT adjustment screw was relocated to the center of the casting in front of the choke housing. An aluminum plug is driven into the access hole for the APT adjustment to prevent tampering. Here, we've raised the plug slightly for the photo.

Buick carburetors would be the factory Pontiac 455 HO carburetors released in 1971. These were the only factory Quadrajets to not use the outer booster rings surrounding the main nozzles. They were not units that had the rings removed; they were actually special castings where the inner booster ring would extend downward nearly to the bottom of the casting.

This was probably done to help improve nozzle efficiency lost from not using the outer rings. However, these units proved to be very inefficient off idle and were discontinued after a single-year production run. As with the 1973 and 1974 Super Duty carbs, the 1971 HO carburetors were produced in very limited quantities and are highly coveted by restorers.

Almost all of the factory high-performance carburetors delivered on the Oldsmobile 442s, Buick GSs, Pontiac Ram Air GTOs, and Firebirds would fall next as the best factory production units.

Here comes the bad news. Most of the carburetors mentioned were part of very limited production runs. Not only does this make them extremely rare, but the purchase price for a decent core would be unreasonably high. At the time of this writing, a decent Pontiac Super Duty core in need of full restoration brings about $600 to $800. Fully restored units have brought in excess of $2,000. The only carburetors mentioned as being among the best Quadrajets that are currently available at a reasonable price are the large-CFM Buick castings. Even those are somewhat difficult to obtain and lack the hot-air-style choke, making them less attractive than other models.

Now that we've outlined the "best" Quadrajets as delivered, let's take a look at the best Quadrajets from a practical and affordable standpoint. In 1975 the Quadrajet underwent a complete makeover. The factory wanted to have closer control of the part-throttle fuel delivery. They basically took the larger castings and revamped them to include an adjustable part throttle (APT) system. Quadrajets produced in 1975 through early 1976 used a unique APT system using a single jet/rod located in the front right corner of the main casting. The screw is accessible after removing a brass cup plug in the air horn.

Starting in 1976 this system was changed slightly, only for the better. What this means to the high-performance enthusiast is that almost all production units made in and after 1975 not only have the larger CFM castings, but also full control of part-throttle A/F mixtures externally. These carburetors have typically been avoided since they were released on pathetically anemic low-compression smog engines produced at the time and were very limited on fuel delivery. In reality, these carburetors have exactly the same potential as the very best factory production carburetors, such as the famous Pontiac Super Duty Quadrajet. They also have several features, due to being of the later, that actually make them superior in several ways when compared to the early carburetors.

Even better news is that these later carburetors were produced in much greater quantity. Since very few folks are looking for them, and there are many of them around, they can be had at bargain basement prices. At the time of this writing, the price range for a late-model front inlet non-ECM core is about $5 to $35.

Inlet Position

The Quadrajet was produced in two different fuel inlet configurations.

All but a very few Chevrolet carburetors are of the side-inlet design. A small number of mid-'70s units were released with the small castings and front inlet on Chevrolet engines. The most common of these were the Chevrolet 454 truck carburetors. These units used front inlet castings of the early design, yet were used in and after 1975. They are indeed special carburetors, but really not all that common and certainly no better than

Shown here are two carburetors from about the same period. The carburetor on the left is a side-inlet Chevrolet model, and the carburetor on the right is a front-inlet Pontiac. Fuel-inlet location can be a major concern for carburetor selection due to interference with items bolted to the engine, such as the thermostat housing.

Early Quadrajets used ⅞-inch -20 fuel-inlet housings. The sealing surface was on the shoulder of the nut. Having the seal behind the threads presented a possibility for material to find its way into the carburetor as the threads began to deteriorate.

Most Chevrolet Quadrajets were side inlets. Some truck units were produced with front inlets, as shown here with this 1976 Chevrolet truck carburetor, number 17056212.

any other pre-1975 front inlet carburetor. Quadrajets were also released with several different fuel-filter housing designs and different inlet thread sizes. Early carburetors have smaller threads with the seal for the housing before the threads.

In the early 1970s this design was replaced with larger threads and a revised seal location before the threads.

The later design is better than earlier designs as it is less likely that small pieces of the threads find their way into the carburetor, and the sealing surface is protected, being well below the surface.

Most of the Buick, Oldsmobile, and Pontiac engines have a Quadrajet with a front inlet. There are exceptions to this rule; the Pontiac OHC 6-cylinder engines used a side-inlet Quadrajet as well as numerous other Buick, Oldsmobile, Pontiac, and Cadillac engines produced in the later '70s through the 1980s.

With very few exceptions, all front-inlet Quadrajets (non ECM) released in and after 1975 have the larger castings. The only exceptions that I've seen were the rare front-inlet Chevrolet carburetors still using the earlier casting designs released in and after 1975. Most of these were used in trucks, Suburbans, and motor homes with big-block engines.

Most of the side-inlet Chevrolet Quadrajets released in and after 1975 also have the larger castings. There are exceptions, and the most reliable way to distinguish between the larger and small castings is by visual observation. Using part numbers to identify CFM is unreliable as there are exceptions to nearly every rule.

Even though one may obtain a smaller CFM casting, they still flow at least 750 cfm when correctly prepared, making them suitable for all but the largest CID street and strip engines.

Choosing an inlet location can be a matter of personal preference aside from convenience. If a modified fuel system is being used with a fuel-pressure regulator mounted on the passenger's side inner fender, then a side inlet may be beneficial as it is in straighter line with the fuel pressure regulator. On certain engines, using a long fuel filter housing combined with a front inlet location can make running a fuel line difficult—the inlet is in close proximity to the thermostat housing. Some enthusiasts are also loyalists and would prefer not to use a front-inlet Quadrajet on their big-block Chevy engine, just as an Oldsmobile 442 owner may not want a side-inlet "Chevy"-style carburetor sitting on his

In the early 1970s, larger 1-inch -20 threads were used and the seal location moved to the front of the housing. This improved sealing and helped prevent any material from the carburetor threads from getting into the fuel bowl.

W-30 455 engine, no matter how good it works! More good news here: the fuel inlet locations have no effect whatsoever on carburetor performance. This opens up the choices for carburetor selection when browsing through the salvage yards and swap meets.

CFM

High on the priority list for carburetor selection should be cubic feet per minute (CFM) airflow capacity. Quadra-

engine is frequently (or constantly in a marine application) heavily loaded. Being able to cruise without applying the secondaries and/or at a lower primary throttle plate angle increases fuel economy considerably.

The primary use of the vehicle and engine size are also factors in carburetor sizing. For "mild" engine combinations and smaller CID, the "750" cfm Quadrajets are more than adequate. If your car was delivered with a factory Quadrajet and made prior to 1975, it

engine combination to at least 650 hp. Since the design of the Quadrajet is basically variable CFM, there are no negatives from using a larger casting on the smaller CID engines. GM installed the smaller carburetors on engines with as little as 305ci, but severely limited the opening angle of the secondary throttle plates and/or secondary air flap for most of these applications.

Linkage

Through the years of production, the Quadrajet was delivered with a broad variety of linkages. Some have hook-ups for transmission downshift or throttle valve (TV) cables, many have a fitting to hook up cruise control. The vast majority have numerous places both above and below the shaft pivot point to effectively install the required fittings to hook up the throttle cable and return spring(s). Some models do not. Many Buick carburetors only have linkage above the center of the primary shaft, as do some later-model carburetors.

This can pose a problem on some applications, particularly if the return spring location is behind the carburetor. Having hookup points available only above the shaft centerline mandates running the return spring(s) to a forward location.

The most accurate method to identify a large primary bore carburetor is by visual observation. Note the "bump" in the large bore casting as indicated above.

jets come in two basic sizes, as long as you're not considering the extremely rare 1971 Pontiac HO castings. For the most part we are looking at either a "750" cfm or "800" cfm carburetor. The smaller castings are sufficient for most engines. Even so, there can be several benefits from using the larger castings. Since the only difference is in the size of the primary bores, less throttle angle is required to obtain any particular vehicle speed when part-throttle cruising using a larger casting. This becomes most beneficial when the Quadrajet is used in a towing or marine application. These applications require large CFM flow and the

probably uses the smaller Quadrajet anyhow. There is really no need to search out one of the larger carburetors in most cases. If you already have one of the larger castings, will be sufficient for any

Three base plates are shown for comparison. The top base plate is from an early Buick carburetor. Note the "hook" and lack of any provisions for return springs or transmission cable attachment points below the centerline. The center base plate is from a Chevrolet carburetor; the bottom is from an Oldsmobile.

Pontiac carburetors from the 1970s used large front vents. The style on the left was used between 1971 and 1974; the model on the right was used from 1975 to 1979. Note that the vents are not in the same exact location and would require air cleaner base modifications if used with anything other than a stock air-cleaner assembly.

The vast majority of linkages contain plenty of potential hookup points for the throttle cable or rod. From the factory, most early carburetors use either a ¼-inch round ball or ³⁄₁₆-inch "post" to attach the throttle cable. Well into the

Several companies offer "kits" to provide attachment points for various throttle cables. Shown here is a kit from Holley.

'70s, the larger ¼-inch posts started showing up. Any of these factory fittings can be removed and a wide variety of aftermarket kits are available for getting the throttle cable hooked up to the carb.

The user must consider the distance up from the center of the primary throttle shaft when hooking up the throttle cable. The farther away the attachment point, the longer the distance of travel is required by the throttle cable or rod to get the carburetor to a full-throttle position. Although a basic point, it is often overlooked as part of the carburetor installation process. The user must also consider the attachment point for return springs. It is highly recommended to use at least two return springs, one inside the other in case one fails. The attachment points should also be tested to make sure that

the springs don't travel too far or far enough to bind up on the linkage. The last item to consider for throttle-cable attachment is the mounting point on the intake. Many aftermarket high-performance intake manifolds are taller or have a different carburetor flange location than the stock intake. If an aftermarket intake has been installed or a spacer placed under the carburetor, the stock cable usually needs to be moved up or relocated to work the throttle linkage effectively.

It is a common practice to use carburetor spacers. We have encountered numerous over-the-counter spacers that interfere with the fast idle cam on certain models of Quadrajets. They simply have to be machined or filed for clearance. Some gaskets also interfere with the fast idle cam and can be easily modified with a heavy pair of shears or tin snips.

We have also encountered minor problems with some factory and low-style air cleaner bases. This problem is usually not discovered until the test drive, so it's a

Modifying Base Plates for ⅛-inch NPT Threads

If your base plate does not have any NPT threads to install a fitting for power brakes, it can be added easily during rebuilding. Very carefully drill through the center of the plate as shown.

Clamp the base plate in a vise with soft jaws. Drill a ⁵⁄₁₆-inch hole through the center of the casting.

Start out with a small bit and increase bit sizes until a ⁵⁄₁₆-inch hole has been drilled. It helps to work the drill in a circle to taper the hole slightly to allow for easy tapping with a ⅛-inch NPT tap as shown.

When using pipe taps, tap only deep enough to leave about three to five threads showing on the tap. Insert the fitting into the hole and test the fit. Threading too deep causes the fitting to bottom out before it is tight. Pipe threads are tapered, and installed fittings seal on the threads to prevent vacuum leaks. The tapped hole should be deep enough to provide approximately five full turns with the fitting to be used. No sealant is required, but a small amount of pipe thread sealant helps prevent leaks if the fit of the parts is less than desirable.

Turn the base plate in the vise with the hole facing up. Carefully start the ⅛-inch NPT tap into the casting. Once the tap gets a couple of threads cut, continue to tap the hole by going ½ turn clockwise, then back ¼ turn counter-clockwise. This clears the thread and keeps the tap from binding. Tap until about three or four threads are showing on the ⅛-inch NPT tap.

good idea to check for full throttle movement without interference before driving the car after a new carburetor is put into service. Starting in 1971, most Pontiac carburetors used a large front vent located in the right front corner of the air horn. This was discontinued around 1980. A large corresponding hole in the air cleaner base is required if using one of these carburetors. We also recommend using a large "O" ring to seal off the opening when the carburetor is in place to keep unfiltered air out of the engine. It is also important to consider that the vent location between 1971–1974 and 1975–1979 Pontiac carburetors is not the same. The later carburetors have the vent slightly more toward the fuel inlet. This can cause minor problems with the orientation of factory "Shaker" and Ram Air pans when swapping carburetors between these years of production.

Vacuum Fittings

Factory production Quadrajets offered a wide variety of vacuum fittings. Almost all models offer both ported and manifold vacuum sources. A high percentage has either a ⅛- or ¼-inch NPT threaded opening in the rear of the base plate. This was usually used for power brakes. It may not be present on some models, and many Buick carburetors have only a small steel tube at this location. If a rear fitting is required for power brakes, it can be easily added during carburetor rebuilding. The factory also installed various Ts or other fittings that offered vacuum for transmission modulators and various other devices.

Most early carburetors have at least one ported vacuum fitting. In most cases the source location is low enough in the base plate to provide a strong vacuum signal right off idle, sufficient for operating a vacuum advance on the distributor. Starting in the early '70s most Quadrajets also used a ported vacuum fitting with a high source location in the base plate. This was to provide a weaker source for operating the EGR valve. This source is usually not strong enough to effectively operate a

Continued on page 51

The factory used various fittings in the rear of the base plate to operate power brakes and other accessories.

Distributor Tuning

Tuning the distributor for a high-performance engine is just as critical as carburetor tuning. The need for a modified or other-than-factory spark advance curve is beneficial, and in most cases mandatory, after engine performance modifications. It is quite common for the engine builder to choose a camshaft with more duration and a tighter lobe separation angle than the stock or factory camshaft. This has the engine not running as well at idle and low engine speeds due to the increased overlap. Camshaft overlap is simply the amount of time in crankshaft degrees that both the intake and exhaust valves are open at the same time as the piston passes TDC on the exhaust/intake stroke. This condition sure makes for a nice-sounding engine, but often brings with it poor idle, rich-smelling exhaust and often engine stalling at low speeds. The first and most immediate cure is to increase the base or initial timing setting. The factory used relatively conservative initial timing settings for most applications. Increasing the ignition timing helps to burn the poorly compressed mixture by igniting it a bit earlier during the compression stroke. In most cases simply increasing the initial timing setting is sufficient to cure our engine-idle and low-speed troubles. Even so, it can cause two additional problems. One problem is difficult hot restarts, and the other is too much total timing caused by the new higher initial timing setting.

Another problem often encountered is that most factory vacuum advance canisters provide a lot of additional advance. It's not uncommon to find 30-degree or even higher vacuum canisters on late-model distributors.

What the user finds is that having that much additional timing is not going to work well with our new engine combination. Everything we do in the quest for

Many stock GM HEI vacuum advance units provide way too much spark advance for high-performance applications. Notice the long slot provided by this unit. It adds an additional 25 degrees of timing when applied, indicated by the last two numbers on the unit.

more power, including increasing airflow, more efficient combustion, and raising both the static and dynamic compression within the engine, requires more conservative ignition timing across the RPM range.

So here we sit, faced with a couple of new problems. Our new engine now requires a higher or more advanced initial timing setting, and it also requires less total timing and less timing added by the vacuum advance. This typically leads to some custom distributor tuning. A high percentage of enthusiasts purchase an aftermarket distributor. There are several very good, reliable self-contained units available that are not only easy to install, but just as easy to tune. Many do not have a vacuum advance canister. We feel that it is best to use a vacuum advance for a street-driven engine. For most applications, it is good for a few more miles per gallon and can also be used to help with idle quality with some combinations. Most of the vacuum advance canisters are adjustable for the apply spring rate, and many are also adjustable for how much additional timing they add when applied.

Ported Vs. Manifold Vacuum

If a vacuum advance distibutor is used, there are two ways to apply the additional advance during engine operation, and which one to use can be a topic of much controversy. What we've seen is that most tuners really don't understand how the vacuum advance works. Most Quadrajet carburetors contain several vacuum sources for operating various factory-installed devices, including the vacuum advance on the distributor. They always offer at least one manifold vacuum source and most have at least one well-located ported or timed vacuum source. We must emphasize *correctly located* "ported source." A correctly located ported source applies full vacuum right off idle, as the source location lies just above the primary throttle plate position at idle. Once the throttle angle is increased to uncover the ported vacuum source, it shows the same vacuum readings as any other source located under the throttle plates.

Ported vacuum does not continue to increase with throttle angle; it falls off at heavy/full throttle just like a manifold vacuum source. The only real difference between the two sources is that the correctly located ported source applies the vacuum advance unit at part throttle only. The manifold source applies the advance at idle, coasting, and part throttle. Any heavy or full-throttle operation does not have the vacuum advance unit applied to either location. So there are no full throttle power differences no matter how the vacuum advance unit is used.

If a ported source is used, there is no additional timing added at idle. This can be beneficial with some combinations. If we add additional timing at idle using a manifold vacuum source, the engine vacuum reading increases and the idle quality improves. At a glance, this looks beneficial. However, if a relatively large camshaft is being used, the engine vacuum can fall off enough when a slight load is placed on the engine that it causes the vacuum advance canister spring to overcome the vacuum signal and retard the timing. This results in the engine stumbling, running rough, and often stalling. Another condition that can occur when using a manifold vacuum source is that the vacuum signal may not fully engage the advance canister or may fall off when a slight load is placed on the engine such as placing the transmission in gear. If the timing retards slightly, this often causes a greater drop in engine RPM and often engine stalling.

Using a ported source has us increase the throttle angle more than if a manifold vacuum source were used. This usually improves idle stability in and out of gear with automatic transmission vehicles, as no signal to the advance is lost when placed in gear. A correctly located ported source applies the advance with the slightest movement of the throttle off idle. This works extremely well for improving throttle response and part-throttle power. There are never any tendencies for the timing to retard at any point from losing signal to the advance. Once the throttle opening is raised enough to uncover the ported vacuum advance source, it has the same vacuum reading as any other location under the throttle plates. This guarantees the exact same function as if a manifold source were used to the advance during all light part-throttle engine operation.

When a manifold vacuum source is used for the advance, the amount of advance supplied by the canister is added at idle and also during coasting. In many cases, adding a lot of additional advance at idle is not

Continued on page 50

The source location for ported vacuum advance is usually a slot cut into the casting just above the throttle plate's position at idle speed. When the throttle is moved off idle, the throttle plates uncover the slot in the castings, providing full engine vacuum to the distributor vacuum advance.

Ported Vs. Manifold Vacuum CONTINUED

Aftermarket vacuum advance units are available in adjustable versions. Shown is a unit that can be custom tuned on the engine by inserting a small Allen wrench into the vacuum supply hole.

Custom distributor tuning often mandates limiting the total spark advance curve. Many factory distributors offer far too much spark advance to operate correctly on highly modified engines. Many do not offer a positive stop for the mechanical advance. When lighter advance springs are used, it is likely that additional timing is added at high RPMs. Here, we have brazed the slot down in this HEI advance mechanism to limit the total timing required for the application and ensure that the timing doesn't continue to increase.

needed. If we choose to use a manifold source with a big cam and our engine has low engine vacuum at idle, it does not typically apply. It also holds the advance in place unless it can be adjusted to a low spring tension or is factory-equipped with a spring that has less tension than the low vacuum reading. With some other combinations, where we already have a relatively smooth idle and high vacuum at idle, adding more initial timing can result in the engine bucking profusely in protest. Adding manifold vacuum advance in these cases typically mandates limiting the amount of advance considerably.

To offset the problems involved with trying to apply the vacuum advance with engines that have low vacuum at idle speed, aftermarket "adjustable" vacuum advance units are available. These allow the tuner to lower the spring tension within the canister so the timing does not fall off at low engine speeds/low vacuum.

We have heard through the years that there are additional benefits to the engine by using a manifold vacuum source for the advance. To date, none of these have been substantiated through our testing, but we still feel it is important to mention them. Increasing the timing setting at idle by using manifold vacuum to the advance has been advocated to improve engine cooling. This could be true, but only if the engine were used at idle and very low engine speeds/load. If using manifold vacuum with your particular engine combination seems to help reduce coolant temperatures when idling for long periods of time, then this is beneficial. Using manifold vacuum for the advance has also been advocated to improve fuel economy. This could be true if the engine were only used at idle speeds. However, in actual operation for most "normal" driving, the throttle position is increased enough to uncover the ported source to the distributor. Our testing has shown no improvement in total fuel economy during

numerous tests when a manifold vacuum source to the advance was used versus using a ported source.

Now the question is "which one to use?" The answer is to use the source that works best for your particular combination. We feel it is better to limit both the mechanical and vacuum advances to the most ideal settings prior to selecting which vacuum source to apply the advance.

Then, through testing, experiment with each source location to determine which one is more beneficial. I've done exactly this with hundreds of vehicles, including my own. With my own vehicle, the only real difference is that the engine slows about 50 more RPM when placed in gear if a manifold vacuum source to the distributor is used. There are no differences anywhere else in terms of improved fuel economy, off-idle throttle response, or engine cooling. At the track the vehicle runs exactly the same ET and MPH if the advance is not hooked up, or hooked to either a manifold or ported vacuum source.

The rear of the base plate was a common point for large vacuum fittings to operate power brakes and other accessories. The top plate has a small pressed-in tube, and the center plate base plate uses ⅛-inch NPT threads, common for carburetors produced through about 1974. The bottom base plate uses ¼-inch NPT threads, typical of many 1975 and later carburetors

Many late-model Quadrajet carburetors use a rear choke pull-off only. The pull-off used two links, one to hold the secondary air flap closed, the other to unload the choke on cold start-ups.

vacuum advance at a low enough throttle opening and should not be confused with the correct ported distributor source. Use a vacuum gauge to make sure that full manifold vacuum is obtained by the ported distributor in source right off idle if it is to be used correctly operating the vacuum advance.

Choke Pull-Offs

The choke pull-off is a very important part of the Quadrajet carburetor. It serves double duty, as it is used to hold the secondary air flap tightly closed and help return it to a closed position during high-vacuum situations. The release time of the pull-off is of importance when modifying the secondaries for a faster opening rate without bog, stumble or hesitation. Some choke pull-off canisters are easy to modify, and others are extremely difficult if not nearly impossible.

The choke pull-off must be a consideration when selecting a carburetor for high performance use. Most Quadrajets, at least those produced through the mid 1970s, use a single front or primary choke pull-off. There are exceptions to this, such as the early model Buick carburetors.

In the mid 1970s, Quadrajets started showing up with a rear or secondary choke pull-off in addition to the primary choke pull-off. This was added to ensure sufficient choke unloading after initial cold engine starts. Around 1980 some units started showing up with the rear choke pull-off only.

These carburetors also used a redesigned air horn that eliminated the slot for the link from the pull-off. The slot, required for correct secondary air-flap

Many different choke pull-offs were used through the years of production. On the left are three common early pull-offs; on the right are three variations of late-style pull-offs made in and after 1975.

Although more common on later units, early Buick carburetors used both primary (right) and secondary (left) choke pull-offs.

This is a typical remanufactured carburetor. Note that all of the plating on the main body and air horn has been removed. Also note the "remanufactured" sticker attached to the main body.

operation, was moved to the rear pull-off instead. The rear choke pull-offs typically have a very slow release time and can be somewhat difficult to modify. They can still be used for high-performance use, but they are not as advantageous as the models that use a primary choke pull-off. They can be converted to the primary choke pull-off. Even so, the later primary choke pull-off designs are so common that choosing one is usually a better path than working with a rear choke pull-off design in the first place.

"Remanufactured" Quadrajets

In the past several decades, quite a few companies have made an effort to rebuild and refurbish used Quadrajet carburetors.

At first, most companies simply rebuilt the carburetors to original specifications and re-sold them as such. As time went by and the number of carburetors that could sell for specific applications became scarce, companies diverted their efforts into producing "generic" carburetors. This was very beneficial from a business standpoint, but pretty much a death sentence to the "remanufactured" carburetors themselves. These companies started batching all of the base plates, main bodies and air horns by group, then installing the same calibrations into each one. This allowed a single part number to cover a much broader range of applications. They didn't have to stock as many parts and the entire process became less complicated. They verified the calibrations with rather sophisticated (and expensive) wet-flow testing equipment. Even so, the vast majority of the samples we have tested are not within calibration. In order to make them more user-friendly over a broader range of applications, they typically have generous idle and off-idle calibrations and have any idle bypass air blocked off internally. I haven't seen a single sample in at least 10 years that works anywhere as well as a correctly rebuilt factory original unit.

Some of the companies also employ a few steps to the rebuilding process that make these carburetors very difficult to revert back to factory specifications. It is usually best to avoid any of the commercially remanufactured carburetors. Even though most can be restored and recalibrated, the process is usually very labor-intensive, and despite your best efforts it can lead to very poor end results

What Quadrajet Do I Use?

We've outlined some of the various models and features found on the Quadrajets. Now, which one is the best for my application? If you already have a Quadrajet carburetor, it may be the best choice. The actual difference between most models is rather slight and nearly any model/year of Quadrajet responds well to the modifications outlined in the Chapter 6. The owner may also be motivated to use the original factory part number if it is still in use, even though it may not be the very best starting point.

If you don't yet have a Quadrajet to work with, I recommend obtaining a

The first float arrangement had the pivot point close to the power piston tower. These early-style carburetors used a large float to keep sufficient leverage on the needle-valve assembly for accurate fuel control. Later designs used smaller floats and improved fulcrum position and make a better starting point for building a high-performance carburetor.

late-model carburetor produced in or after 1976. Avoid the early-style APT units and those containing the auxiliary power piston.

Your carb should have your desired fuel inlet location. Any of the front-inlet units already have the larger castings, as do most of the side-inlet models. Any side-

Several indicators tell us we have selected a good late-style core for building our high-performance Quadrajet. Note the aluminum plug over the APT adjustment and raised "tower" in the right front corner of the air horn.

inlet model chosen should be visually inspected to make sure it has the larger primary bores if a larger CFM carburetor is desired. The carburetor chosen should have the linkage desired, or at least enough room to effectively hook up the throttle cable with minimal modifications, and a location for the return spring and throttle kick-down linkage if needed. The vast majority of the carburetors produced in the later years of production are generous with both throttle linkage hookups and kick-down linkage, or at least a place to attach it

We also advocate using the later carburetors because they have several very nice features and are typically not nearly as worn or weathered as the older carburetors. The vast majority of the carburetors from the 1960s show considerable porosity in the castings, warping across the front, stripped threads, and they nearly always leak at the bottom plugs. Most of these shortcomings can be corrected or repaired, but they typically require more labor time to do so.

For full race applications, we do not recommend using models made prior to

In the early 1980s additional screws were added to hold the air horn to the main body. Instead of one single retaining screw in the casting directly in front of the vent, five retaining screws were used, as noted in red.

1969, or any later design that still used the early-style float/fulcrum arrangement. They can be made to work, but the lack of leverage on the fuel-inlet needle and large float required make them harder to keep full on hard runs. Later units that used the revised float fulcrum arrangement and smaller float prove to be a better choice in the long run.

Additional advantages to using 1976 or later units are that they are much more readily available at more reasonable prices. Rochester continued to improve the design right up until the very last units were produced. The casting materials are denser and less prone to warping. They added additional screws in the base plate and air horn to keep the gaskets from leaking.

Most are the larger castings and all use the smaller float and improved fulcrum position. Most had the choke on the

In the late 1970s an additional attaching screw was added to the base plate.

carburetor for easy conversion to electric chokes, and many already have electric chokes if produced in and after about 1980. They also have APT, giving the user full controls of the part-throttle A/F mixture.

The bottom line is that there is no performance difference between a fully prepared 1976 or newer large casting and the 1973–1974 Pontiac Super Duty carburetors, which are among the very best high-performance carburetors ever produced. The newer units actually have a few advantages over the SD carburetors as described above. Now that we have a good idea of what Quadrajets to start with, let's find out how to make it work to our expectations.

TOOLS AND SAFETY

Rebuilding and modifying the Quadrajet carburetor is not incredibly difficult. It requires some basic hand tools and other equipment recommended in this chapter. Many of the items mentioned are nothing more than standard hand tools normally found in most home tool kits or mechanic's sets. Very few "special" tools are needed. In this chapter we mention many different tools and specifically what they are used for. Some of these tools have been modified slightly for a specific use. We also go into greater detail with a few of the more complicated machining processes involved for those with the access to these tools and the skill to do the work.

Safety Equipment

Safety equipment can be more important than getting the actual work done. Before any job can be performed, you should outline the procedures needed to accomplish the task and the required safety gear to perform the work. First and foremost on our list is hearing protection.

I know that a lot of you are thinking, "Why in the world would I need hearing protection to build a carburetor?" Many of you will be using high-pressure compressed air and blowing it into tiny passages. The high-frequency sound produced could affect your hearing. Some tapping, pounding and banging with hammers and punches is required for different operations. Even though the noises produced are not always sharp or extremely loud, the exposure is cumulative and over time could produce hearing loss. As prevention, purchase and wear hearing protection anytime you work in an environment where you may be exposed to loud noises.

Working with carburetors also requires the use of solvents, most specifically carburetor cleaners and brake cleaners. These chemicals are not nearly

Regardless of what type of hearing protection is preferred, it should be used any time high-pressure compressed air is used. The high-pitched sound created by forcing high-pressure air into tiny carburetor passages can cause permanent damage or hearing loss.

as harsh today as in years past, but they still contain warning labels regarding contact to the skin and inhalation hazards. We use and highly recommend chemical-resistant gloves. Solvents should be used in a well-ventilated area, as high concentrations can produce an explosive hazard and inhalation dangers.

A Place to Work

Rebuilding and modifying your Quadrajet carburetor does not require a large work area or elaborate shop set-up. I built many hundreds of carburetors on a single 2- x 4-foot work table with two vises, one mounted near each end.

Nothing too fancy is required, but you need a suitable place to work on the carburetor. The area needs to be clean and free from airborne dust/dirt. The area should be well ventilated and preferably detached from the house, or at least isolated from any room containing a furnace or appliance with a pilot light.

The work area could be as elaborate as a full-size workbench or as small as a tray table. The more room you have, the less complicated things usually become.

A small workbench is all you need for carburetor rebuilding. It's handy to have a vise with soft jaws close by for holding aluminum carburetor parts. Soft jaws can be fabricated from aluminum angle or pieces of wood.

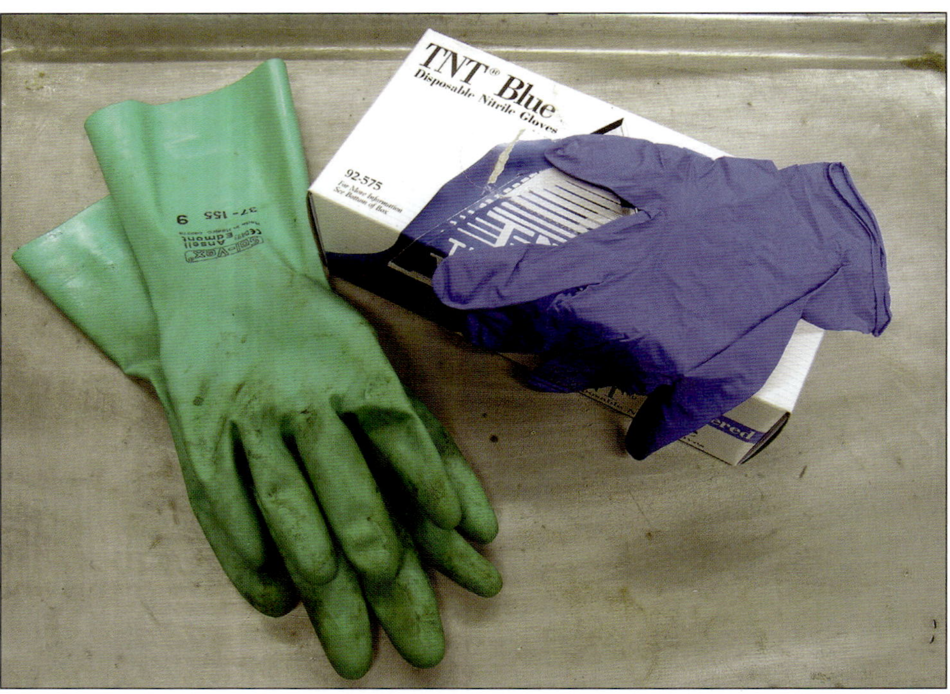

Solvent-resistant gloves protect hands from harsh chemicals. Nitrile gloves are far more chemical-resistant than medical.

Eye protection is highly recommended. Use of chemical solvents, high-pressure air, drilling, grinding, sanding, etc., dramatically increases the risk of flying particles and projectiles. Even the tiniest piece of metal in your eye can require a trip to the doctor to get it removed. I can't think of anything that burns more than carburetor cleaner in your eye(s). Get eye protection! Those with side shields are highly recommended.

Several types of eye protection are shown. Side shields are recommended, especially for any sanding/grinding operations.

A small section cut from a piece of soft pine makes an excellent carburetor stand. It also protects the soft metal during staking procedures.

We've seen some builders develop elaborate carburetor stands to hold the carb. Here, we prefer to cut a 4- to 5-inch piece of wood from a 2 x 4 or 2 x 6.

This allows you to place the carburetor on a flat surface without the linkage resting on the workbench or tray. It is actually quite easy to bend the throttle shaft, so it's best not to put any unneeded pressure on it when removing the air-horn retaining screws. The small wooden block also protects the main components from damage during staking operations, driving in tubes, etc. It is also beneficial to have some sort of tray under the carburetor during the rebuilding process. A large cookie sheet or pizza pan works well. The pan keeps all the parts in one location and from rolling off the bench onto the floor.

Basic Hand Tools

Most early Quadrajets used slotted-head screws to retain most of the components. These require several different sized flat-tip screwdrivers to remove them. The bottom screws that attach the base plate to the main casting were #2 Phillips head. A standard flat tip removes the

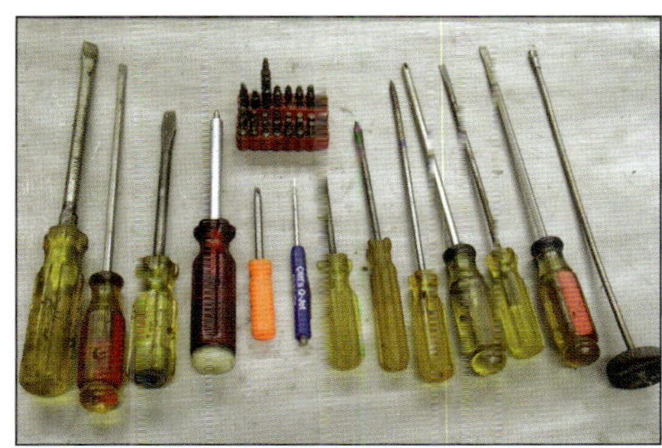

Several different-size screwdrivers are needed. Having a magnetic screwdriver with removable bits simplifies things. The magnetic screwdrivers are also beneficial for removing and installing screws in hard-to-reach places.

screws that attach the air horn. Make sure the screwdriver selected has a tight fit to avoid damage to the screw heads. The two screws within the choke housing may require a screwdriver with a slightly smaller tip. Removing the fuel-inlet seat requires a large flat-tip screwdriver at least $7/16$ inch wide. It is recommended that the tip be hand ground or filed for an exact fit as the soft brass inlet seats are easily damaged or "dog eared" by poor-fitting screwdrivers. Large flat-tip screwdrivers also have a wider section that may hit and even break off the small posts on the '67–'68 carburetors, so be extremely careful when removing the fuel-inlet seat. It is recom-

These large baking pans make a great work area. They are large enough to hold all the parts of the carburetor without piling them up. They also keep items from finding their way off the bench and onto the floor!

A large custom ground flat-tip screwdriver is helpful to remove and install fuel-inlet seats. The sides should be ground down so they don't interfere with removing the seat on the early-style carburetors as shown.

mended that your "custom" fuel-inlet seat screwdriver have the wide section narrowed or removed to prevent damage to the carburetor.

Very late carburetors use Torx head screws in lieu of flat or Phillips head. Several different sizes were used, so a magnetic-tip screwdriver with an assortment of removable tips including Torx tips is handy for later carburetors. Many sets do not contain the tiny Torx tip needed to remove the small #3-48 screws that retain the secondary metering rod hanger and the throttle plates to the shafts. This tip can be purchased separately. Magnetic-tip screwdrivers also help to lift screws removed from the main body such as the one that retains the checkball and the two inside the choke housing.

The jets are made from brass and are somewhat soft. It is highly recommended that a custom wide-blade flat-tip screwdriver be used to remove them. It should be ground or hand filed for an exact fit. Some jets can be very difficult to remove, especially if the carburetor has had any water in the bowl for any length of time. A screwdriver with a good fit in the jets combined with a quick tap from a small hammer removes the most stubborn jets without damage.

Some early Quadrajets used a tiny adjustment screw under an aluminum cap in the center of the base plate between the idle mixture screws. This screw can be very difficult to remove and requires making a small custom flat-tip screwdriver that has been narrowed enough to fit the threaded hole for the screw.

We also need a tool to remove the 1976 and later APT screws located in front of the power piston. Grinding a slot into the tip of a 5/16-inch hardened bolt makes an excellent tool to remove or adjust the late-style APT screws. This same tool can be used to adjust the late-style idle mixture screws.

A small pair of point-nose pliers is needed to remove any clips that are

Even the tiny #3-48 screws were replaced with Torx head fasteners in later models. They require very tiny #2 bits that often have to be purchased separately.

A magnetic-tip screwdriver and bit assortment comes in handy when working with carburetors. Torx head fasteners replaced the early flat-tip and Phillips-head fasteners in the later years of production.

Cutting a slot into a hardened 5/16-inch bolt makes a nice tool to adjust late-style APT screws. The same tool also adjusts some late-model idle-mixture screws.

Use a small flat-faced tapered punch to stake the collar in place to retain the power piston assembly. This is much more effective and less damaging to the carburetor than using a wide flat-tip screwdriver.

Several types of pliers are needed for various steps in carburetor rebuilding. The tiny needle-nose pliers are great for removing and installing small clips in hard-to-reach places.

Early APT carburetors used a tiny flat-tip screw recessed into a narrow passage in the base plate. Removal of this screw requires grinding a custom-tip screwdriver with narrow sides as shown.

to drive the roll pin inward to remove the accelerator pump arm on most 1970 and newer carburetors. Punches also come in handy for other operations.

Precision drilling requires the use of a center punch to get the drill bit started in the correct place. Several items also require "staking."

Staking is a procedure used to move a small amount of material in order to keep something in place. The most common place for staking is at the top of the holding links in place for the choke, accelerator pump lever, and the choke pull-off link where it attaches to the choke pull-off.

Most early-model carburetors use the small clips; later units did not and require slightly different procedures to remove them. A small punch is needed

A variety of punches are needed for driving out roll pins and staking procedures.

power piston tower to keep the collar just above the piston seating during carburetor assembly. A ¼-inch straight wall punch also comes in handy for removing secondary fuel-supply nozzles without damaging them.

A small machinist's hammer is needed for several operations, including

A vise outfitted with a set of soft jaws comes in handy for holding idle tubes, base plates, throttle shafts, and other small items that may need some repair work. Soft jaws can be purchased separately made of either brass or aluminum. It's just as easy to make them from a small section of aluminum angle. The

vise chosen has to be large enough to hold a base plate and small enough for fine work like holding idle tubes, etc. Here we use two different vises, one for small repair work and the other for firmly clamping items we're going to weld, pound, and file or try to straighten.

Drill Bits

A complete drill bit assortment is needed, as we are making many modifications to the carburetor as far as air and fuel delivery are concerned. Precision drill bits are available in numbered kits. The tiny bits are difficult to hold and most of the smaller ones do not fit in a drill chuck. A pin vise is required to hold them and all drilling with tiny bits should always be done by hand as they are easily broken.

Most of the holes to be drilled are drilled by hand using great care not to break or bend the tiny drill bits. Since most of the material to be drilled is rather soft, being made of aluminum or brass, it drills quite easily by hand with light pressure. Drilling by hand also has a great advantage over using a power drill. When drilling by hand, you can feel the drill bit load up with chips or bind up just as it clears through the back

Continued on page 62

A small machinist's or ball peen hammer is handy for light-duty staking procedures. Here it is being used to drive a bronze primary shaft bushing into place.

driving small punches to remove roll pins and to stake any small screws replaced on throttle shafts so they do not come loose and fall into the engine. Installation tools for bronze shaft bushings and secondary fuel nozzles can be made from bolts with a long shank and nut tightened in place. The flat portion of the nut acts as a driving surface while the threaded portion of the bolt guides the part being driven into place.

Our bench vise should be lined with soft jaws to protect carburetor parts. A couple of pieces of aluminum angle make great vise liners. Small blocks of wood can also be used.

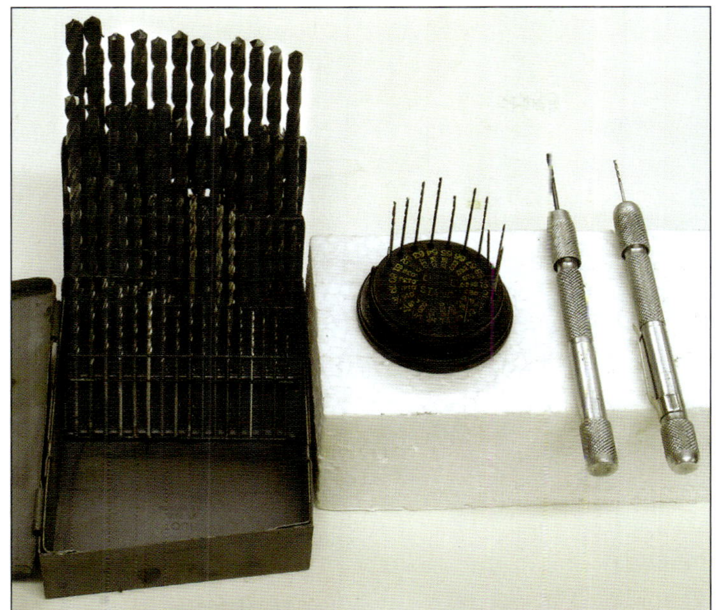

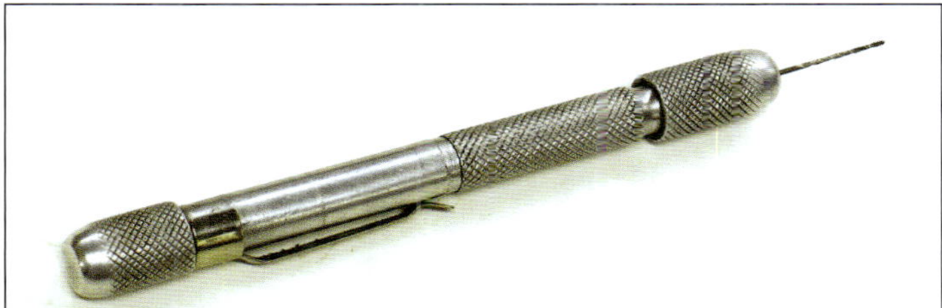

Precision drilling requires the use of tiny numbered drill bits. The tiny bits are difficult to hold. A pin vise is used to hold them and for drilling operations. The tiny bits are very delicate and easily broken.

Some of the larger bits do not fit the drill chuck. The pin vise can be inserted into the drill chuck for some drilling operations that are too difficult to do by hand.

Tapping Threads

For most rebuilding it is not necessary to tap threads in the main body and base plate. On occasion one might encounter damaged threads or get small particles of sand or blasting media into the threads during cleaning operations. The following sizes may be needed: #3-48, #6-32, #8-32 and #10-32. Use of taps this small requires extra care as they snap off quite easily if they become jammed. The correct procedure for tapping threads is to turn the tap $1/2$ turn clockwise, then back $1/4$ turn. Continue this procedure until the tap reaches the bottom of the threads. If you start to encounter a lot of resistance to turning the tap, it never hurts to remove the tap and blow out all the chips with compressed air. Clean the tap with a soft-bristle wire-brush first, blow it off with air, and continue the tapping procedure. Taps come in different styles. Some have a long tapered section and are called starting taps. Others have threads right up to the end and are called bottoming taps. Using taps requires extra care. Even so, they do break off on occasion, usually when you least expect them to. They can be very difficult to remove and are too hard to drill

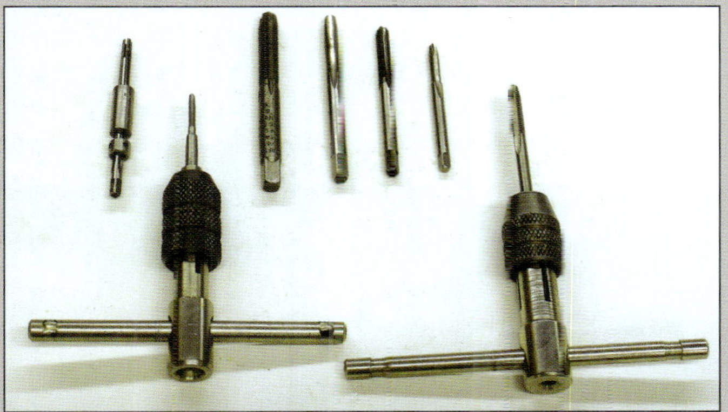

Small taps are used to chase threads to clean out dirt and debris. They require special holders. Use extreme care when using taps as they are extremely hard, which makes them brittle and easily broken. A tap extractor, shown on the left, is used to remove a broken tap. Taps are too hard to drill out with drill bits.

out with most drill bits. Tap extractors are available for most sizes and should be the first method used to remove them, as it prevents damage to the carburetor.

side of the material being drilled. This is when a drill bit is most likely going to break. If it binds up we can reverse the direction of the drill bit and then go back in the clockwise direction while reducing pressure a bit to keep from snapping off the bit. Drill bits are made of rather hard material. It's always best to take extra time to avoid breaking one off, rather than breaking one, then having to sit there wondering how we're going to remove it.

A small electric drill helps with drilling larger holes. Larger drill bits are far less likely to break off than the smaller ones. Most often they slip in the chuck when we get into trouble rather than snap off in the material being drilled. Battery-powered drills are quite handy and don't require dragging a power cord around the shop. A small inexpensive drill is fine. Just make sure the chuck is small enough to hold bits down to about .050 inch. For drilling larger holes, the pin vise can be clamped in the drill chuck.

Sand/Media Blasting

Small sand and media blasting cabinets are available at reasonable prices. For those that are intent on having the small parts cleaned and possibly re-plated, blasting with fine abrasives proves to be the best method. The main body can also be blasted to remove stubborn scale and deposits both inside and out. Any steel parts that are bead blasted rust quickly, as this process removes all of the factory plating. These parts should be re-plated or painted within a few days of the blasting process to prevent rust and corrosion. There are quite a few aerosol paints on the market that can be used to paint external parts on the car-

Bead blasting removes any protective coating on carburetor parts. Several companies offer paint-to-coat castings and small parts to prevent rust and corrosion. They should only be used on externally exposed parts, never on anything internal or that comes into direct contact with fuel.

For the serious carburetor rebuilder! Small bench-top sand/media-blasting cabinets quickly remove rust and corrosion from small parts and main castings.

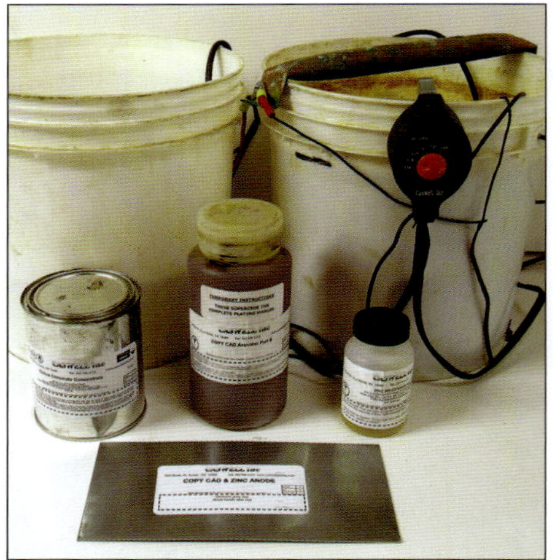

Electroplating kits are available for plating small parts. The kit shown is available from Caswell. They are inexpensive and easy to use. With a little practice you can get professional results at home. The plating process protects exposed metal parts from rust and corrosion for many years.

Preparing the Carb for Media Blasting

If you plan on blasting the main body of the carburetor, take the extra time to install screws into all of the threaded holes.

Sand, bead, or media blasting quickly removes rust and corrosion from metal parts. If the main body is being blasted, installing a few used screws into any threaded holes keeps the abrasive particles from getting trapped in the threads.

This keeps small particles of sand or abrasive media from getting trapped in the threads. This saves time later when the carburetor is assembled, as the smallest particle of sand prevents screws from being tightened up, thus requiring additional cleaning/tapping of the threads.

It is essential that all the small particles be removed with solvent and high-pressure air or they will cause carburetor problems after assembly. Blasting parts not only cleans them, but also removes any protective coating from them. This usually causes them to quickly rust (steel parts) or oxidize (aluminum parts). At this point some sort of coating should be applied to the parts or they should be re-plated.

buretor. Some are specifically designed to provide the original cast-finish look. Similar products are available for the main body and air horn to restore the original cadmium look to the castings.

The small steel parts of the carburetor can also be electro-plated. Several companies offer small "do it yourself" kits that produce professional results. These kits are safe and easy to use and quickly restore the original finish to metal parts.

Air Tools

Air tools are one of life's greatest inventions. Small high-speed air grinders, available in straight and 90-degree shafts, reduce the labor involved in carburetor rebuilding by hours. I still remember spending countless hours removing stubborn gasket material from engine parts during my first overhaul. One of the greatest little inventions is abrasive-sanding/polishing disks. They quickly remove the most stubborn gaskets without damaging the base metal. There are many different grades available, from extra fine to coarse. For aluminum and soft metals, the fine and extra fine disks work best. The coarse materials can gouge the base metal and leave "tracks" in it that need to be polished out.

Air grinders are also needed to grind off the backed staked portion of any screws that retain components to the primary and secondary choke, and secondary air flap shaft. Failure to remove the staked portion of the screws prior to removal often results in the screws breaking off. Since these screws are hardened, removing them is very difficult and often results in ruining the shaft.

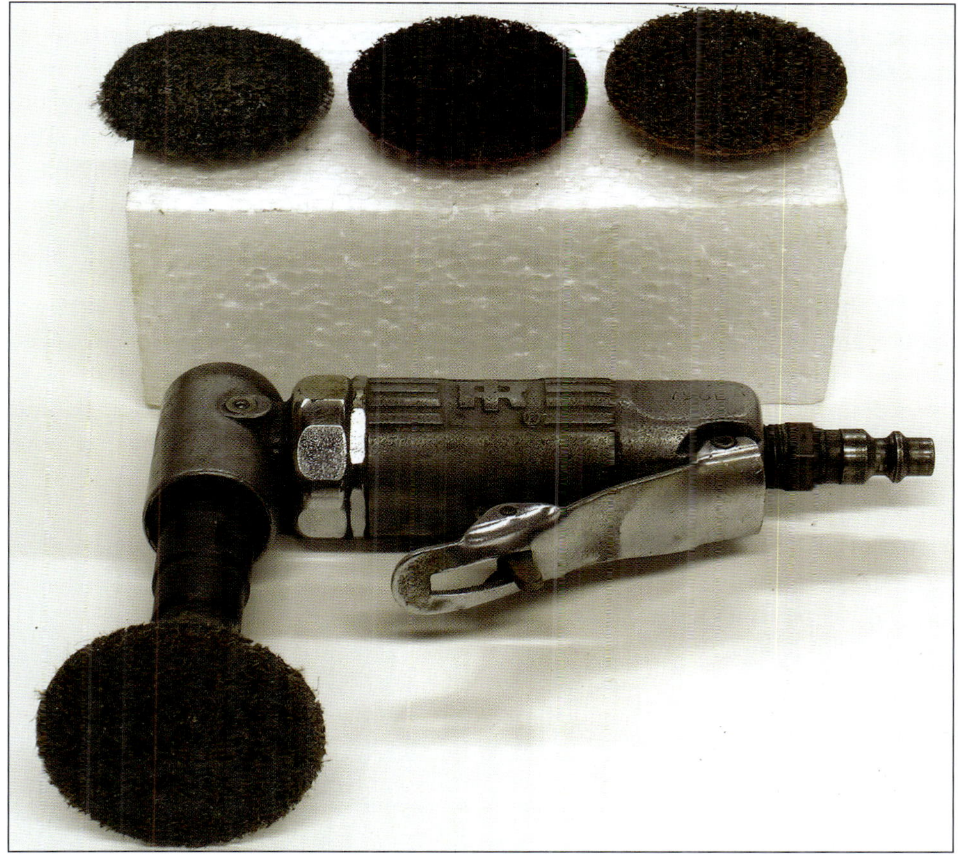

An angle grinder can be set up to use abrasive disks for gasket removal. The fine disk, shown on the left, remove gaskets from soft aluminum surfaces without damage.

The bottom portions of the throttle-plate retaining screws are staked at the factory. This keeps them from working loose and falling into the engine, but makes removing them nearly impossible without breaking them off in the shafts. Grinding the staked portion with an air grinder armed with a carbide cutter allows for easy removal of the screws.

Even the best carburetor cleaners do not remove all the grease and grime from the castings. An assortment of soft-bristle brushes helps remove stubborn deposits without damaging the carburetor surfaces.

Propane Torch/Oxygen Acetylene Welder

Occasionally you encounter a fastener that does not turn, or a frozen/stuck throttle shaft. It's rare to ever be able to remove an early-style APT screw from a base plate without applying some heat to loosen it up. Some idle tubes are stubborn enough that heating the casting around them for about 30 to 35 seconds makes removal much easier. It's also beneficial to warm the main castings a bit before applying epoxy to the main well plugs. A small inexpensive propane torch proves sufficient for all but the most stubborn screws.

If a main shaft has been exposed to the weather to a point where gently heating the casting doesn't free it up, it's probably better to obtain another base plate, rather than spending a lot of time trying to salvage it. It also helps to have a small bucket of kerosene or diesel fuel for soaking heavily oxidized parts. The light fuel oils penetrate deeply into the materials to aid in freeing them up. Several companies also market excellent products to help remove frozen or stuck fasteners. They work as advertised.

Carburetor Cleaners and Solvents

A wide variety of carburetor cleaners and solvents are available to help remove dirt, grease, oil, and grime from the carburetor. Prior to exposing the carburetor to an array of solvents, it should be thoroughly disassembled. Several companies market special cans or small buckets of solvent and a basket specifically designed for soaking the main components to remove large deposits of dirt and grime. The chemicals currently available are much safer than those available in the past. The added safety also means they don't work quite as well. Expect to use a soft-bristle brush for heavy deposits.

Carburetor cleaners also help to wash away some of the grime. Some of these products leave behind an oil-based

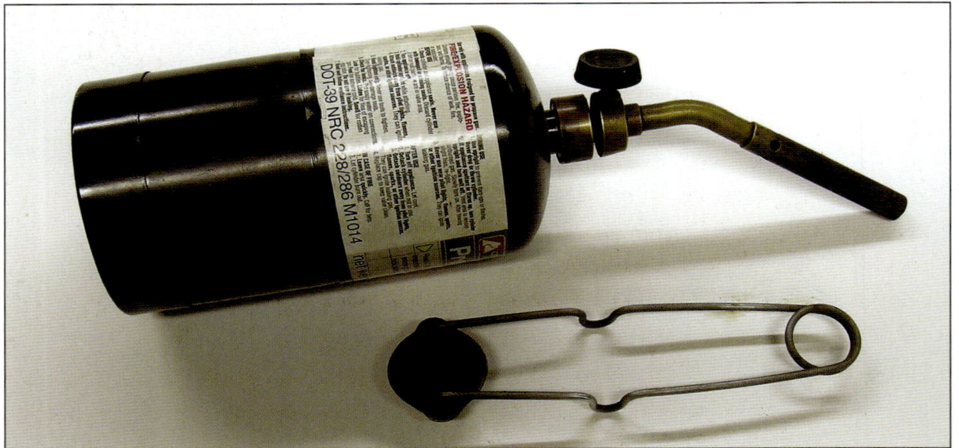

A propane torch can be used to loosen up the idle tubes in the main casting.

Carburetor and brake cleaners are available in aerosol cans. Some carburetor cleaners leave behind an oily coating. Brake cleaners are designed to remove oil and grease and do not leave behind any residue. They can be used for final cleaning of the parts and should be used prior to applying any coatings or plating.

Through the years of production several styles of idle-mixture screws were used. Early units had the screws readily exposed for easy adjustment with a flat-tip screwdriver or ¼-inch drive socket. As engines and engine compartments became more complicated, access to these screws became more difficult. Several companies market special flexible-shaft adjustment tools specifically for adjusting the idle-mixture screws.

Idle mixture screws also come in a variety of shapes and sizes. Most early screws can be adjusted with a flat-tip screwdriver or long ¼-inch socket mounted on a nut driver.

In later years, the factory made several attempts to prevent tampering. They not only installed caps on or over idle-mixture screws, but they also required "special" tools to turn them. The most common idle-mixture screw encountered on late-style carburetors has two flats. Long, flexible shaft tools come in handy for adjusting these idle-mixture screws.

protective coating. It may be necessary to use brake cleaner or electrical component cleaners to remove any oil in order to apply epoxy to the bottom plugs. Many of these cleaners pose inhalation hazards. It is always best to work in a well-ventilated area and wear the appropriate protective gear.

"Special Tools"

Several items on certain carburetors require either special tools or special procedures for removal, adjusting, etc. Idle-mixture screws are one of these items.

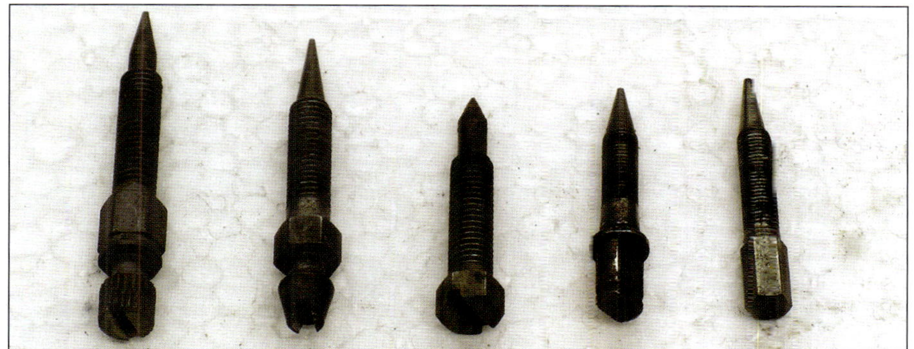

Several different varieties of idle-mixture screws were used through the years of production. Late-model carburetors started showing up with screws requiring special tools to adjust them, as seen on the two samples on the right.

We can make up our own adjustment tool for idle-mixture screws. Early idle-mixture screws were easy to access, and many have a ¼-inch nut in addition to a slot for a flat-tip screwdriver. Using a ¼-inch drive socket with a universal helps access difficult-to-get-to mixture screws.

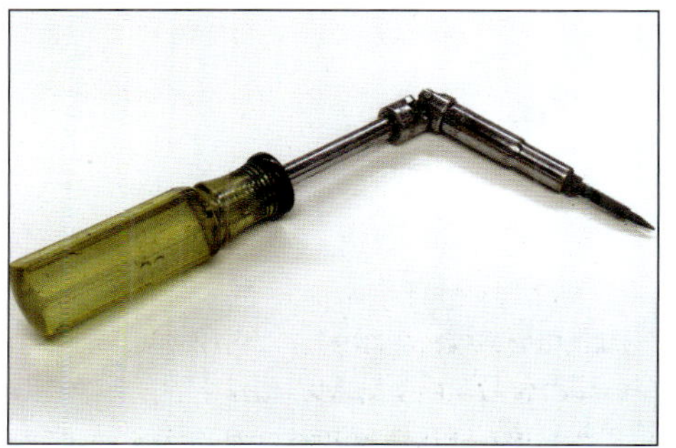

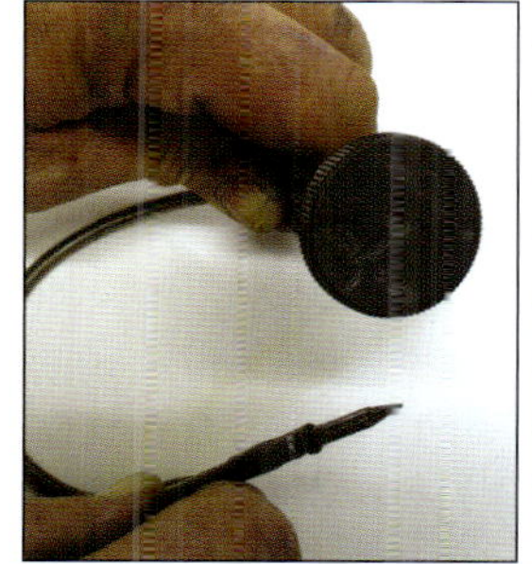

An idle-mixture adjustment tool with a flexible shaft helps get to difficult-to-access adjustment screws.

Many Quadrajets also have adjustments for fine part-throttle metering control or adjustable part throttle (APT). Early APT carburetors only require a very small flat-tip screwdriver to turn the adjustment screw in the base plate. Later models with the APT under the access hole in the air horn require a slotted tool or modification to the adjustment screw itself. The brass adjustment screw can be slotted so it can be turned with a small flat-tip screwdriver.

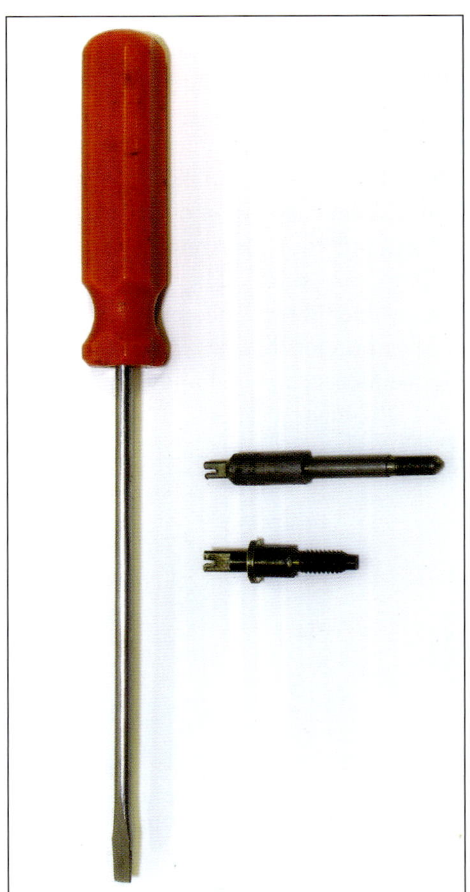

Either style APT adjustment screw can be adjusted with a small flat-tip screwdriver after slotting it with a fine hacksaw blade.

Most Quadrajet primary throttle shafts show excessive wear in the base plate. The installation of bronze shaft bushings is mandatory to ensure that all incoming air goes through the carburetor. Several kits are available to drill or ream

Slotting Late-Style Mixture Screws

Late-style idle-mixture screws can be modified so they can be adjusted with a small flat-tip screwdriver. Grind a small amount from the surface as it is hardened.

Gently clamp the idle mixture screw in a soft-jawed vise. Using a hacksaw with a fine blade, very carefully slot the screw in the center. This allows for easy adjustment with a flat-tip screwdriver once the carburetor is placed in service. An alternate method for adjusting the late-style idle-mixture screws is to make a custom tool from a small bolt.

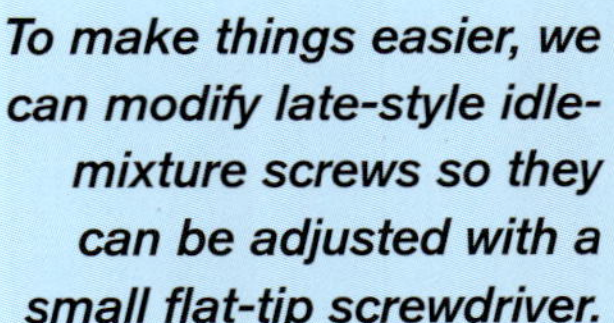

To make things easier, we can modify late-style idle-mixture screws so they can be adjusted with a small flat-tip screwdriver.

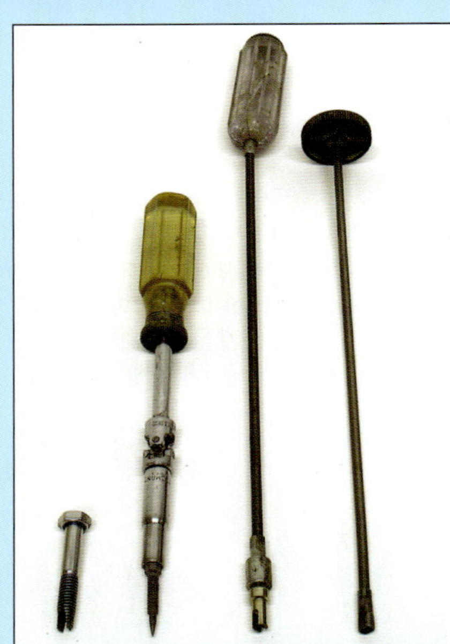

The slotted bolt on the left also doubles as an idle-mixture screw-adjustment tool for late-model carburetors. Two variations of the flexible shaft adjustment tools are shown on the right. The longer of the two has removable tips.

Our basic throttle-shaft bushing installation kit contains a modified self-guiding drill bit, #3-48 shaft-throttle plate screws, bushings, and a small tube of Loctite.

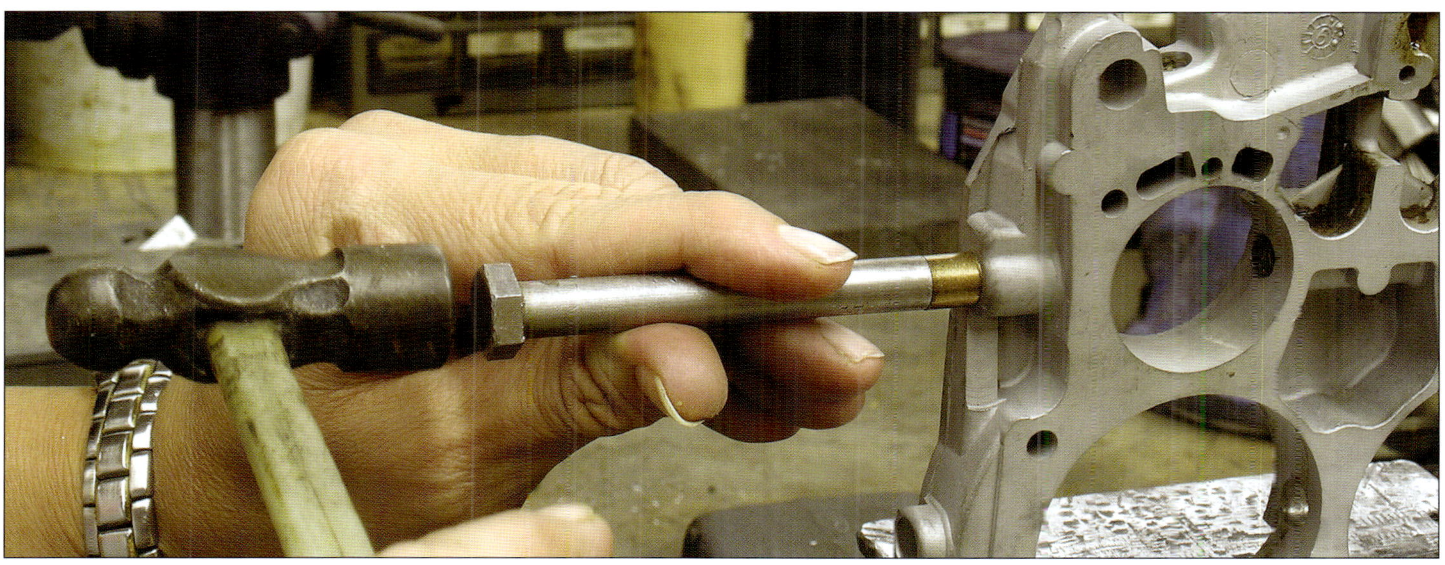

Warped base plates and air horns can be sanded flat on a bench-mounted belt sander.

Aftermarket self-tapping fuel filter housings for stripped out threads.

the throttle shaft and install bushings. We have developed our own kit. A special self-guiding drill bit is used to open up the base plate. Self-lubricating bronze bushings are installed, and the throttle shaft is replaced. The bushings last just about forever and eliminate any potential problems from excessive primary shaft clearance.

A small inexpensive bench-mounted belt sander proves useful for the serious hobbyist. It can be used to clean/straighten base plates as well as warped air horns. The Quadrajet by design can tolerate considerable warping at the air horn and continue to function. Some air horns are high in the front/center and leak at the gasket. Running them carefully across the belt sander removes some of the material at the outer corners where the front bolts go through. This allows the air horn to conform to the main body and prevent leaks at the top gasket.

It is also possible to have the main body machined flat. Keep in mind that this lowers the power piston the same amount that is taken off the main body and can affect carburetor function. We reserve this sort of repair for only the rarest and most valuable carburetors. In most cases it is more cost-effective to simply get a better core in the first place.

On occasion we may encounter stripped-out threads when working with a carburetor. The most common place to find stripped threads is in the fuel filter housing. During the life of the carburetor, the fuel filter may have been changed numerous times. Oversized self-tapping fittings were often used in lieu of thread repair to the castings.

The best method to restore threads for the fuel-filter housing is to have a Heli-Coil installed. Since the fuel-filter housing uses a seal where it contacts the main body of the carburetor, it must be in exact alignment with the main body. This requires precision when cutting the threads for the Heli-Coil insert. The Heli-Coil kit for this procedure is also somewhat expensive. This repair is best left to those with extensive experience. Thread-repair kits are available for every thread size found on the Quadrajet carburetor.

Once in a while we encounter stripped-out threads in the main body. Installing a Heli-Coil or other thread insert is a good method of repair. Another

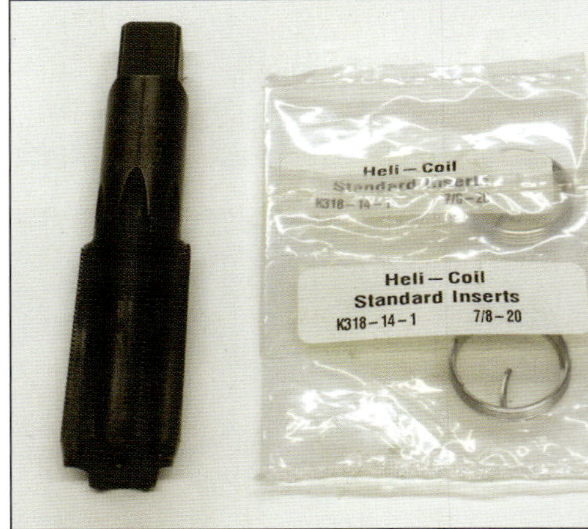

The correct repair for stripped-out fuel-filter housing threads is to install a Heli-Coil insert. Due to the cost of the tap and inserts, and the precision involved during this procedure, it may be best to send the casting out for this repair.

alternative is to find a slightly longer fastener to engage good threads below the stripped-out ones. The air-horn retaining screws are a good example. Early carburetors used #10-32 screws that were 1 inch long to retain the front portion of the air horn. Most of the holes accept #10-32 screw up to one and a half inches long. The longer screws may offer an option for stripped-out threads in lieu of installing thread-repair inserts.

Measuring Devices

During the rebuilding process we must make very precise settings to the carburetor, such as setting the throttle plates to exactly 90 degrees, and adjusting the secondary air-flap angle. A small machinist ruler proves useful for both.

Small metal rulers are also available with a stop. They can be locked in place and used to set the float level more precisely than a ruler or paper scale that come in many carburetor kits.

We also need to measure the diameter of both primary and secondary metering rods. Even though these parts are often stamped from the factory, they may not be legible. The only way to know exactly what diameter they are is by precise measurement.

We have outlined the tools, equipment and supplies needed for rebuilding our carburetor. During the rebuilding procedure, keep in mind that dirt and debris are the biggest enemies of the carburetor. The smallest particle lodged in the wrong passage can cause operating issues when the carburetor is placed in service. Always take the time and precautions to keep the carburetor and the work area clean at all times. Cover the carburetor with a soft cloth or plastic bag between rebuilding sessions.

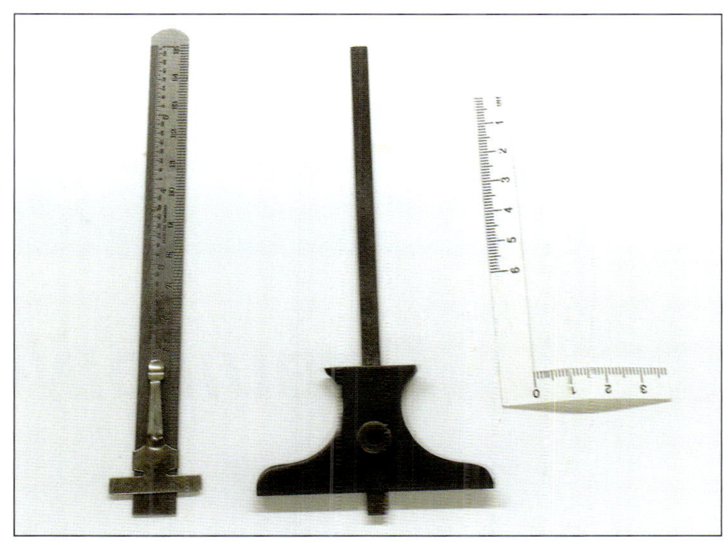
Some metal rulers are available with a slide top. They work great for getting an accurate float setting.

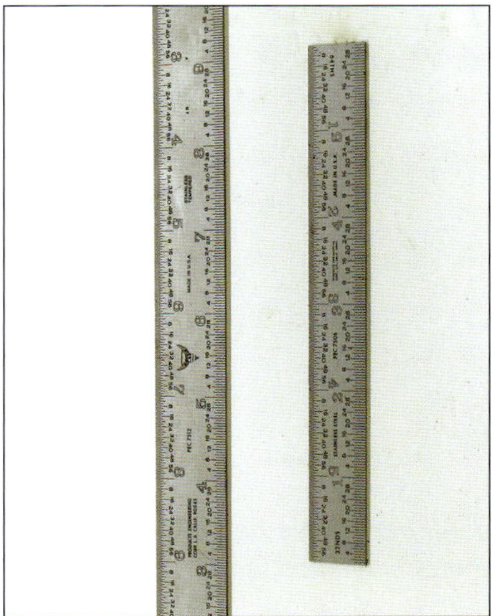
Steel machinist rulers are more accurate and rigid than wooden or plastic rulers. In addition to measuring distance, they can also be used to determine if the throttle plates are set to exactly 90 degrees.

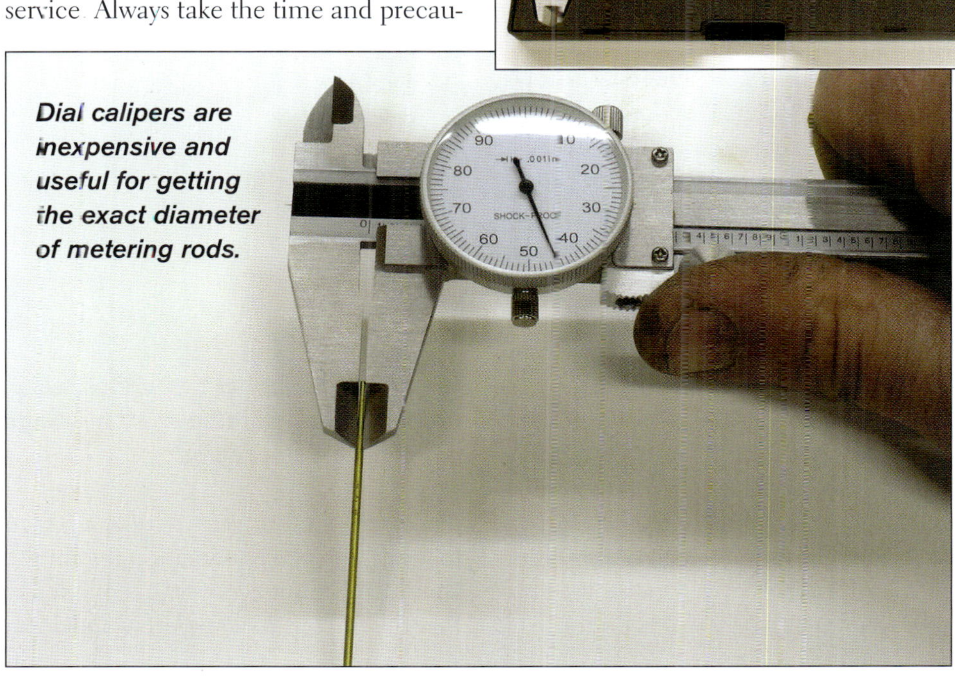
Dial calipers are inexpensive and useful for getting the exact diameter of metering rods.

REBUILDING THE CARB

No amount of effort spent on building a custom high-performance Quadrajet carburetor can be successful unless we are working with a correctly rebuilt carburetor. Over the years I've come to hate the word "rebuilt." Most things I see that are "rebuilt," especially carburetors, were probably better off slam worn-out than after they were finished! This simply comes from lack of knowledge rather than lack of skills. Armed with accurate and useable information, we can make sure that your Quadrajet carburetor is correctly rebuilt, so it functions as it was originally intended. Working with a unit that is up to par, we can go in and make metering and calibration changes to get the desired and predicted results.

Most Quadrajets have experienced some wear through many years of use and countless heating-up and cooling-down cycles in the engine compartment. This may leave some units in rather poor overall condition. The decision to work on your current unit or select another core should be made after careful disassembly and inspection of the castings. Before going into specific rebuilding procedures we go over a few of the common problem areas. Once you have inspected the carburetor and decided that it is in suitable condition to rebuild, you can continue on with confidence that the final product works as expected.

Here is a late-model carburetor ready to go. All parts have been plated, bronze primary throttle shaft bushings have been installed, and the bottom plugs have been sealed.

Bottom Plugs

The design of the Quadrajet is excellent. There are no places for the

Very early carburetors used pressed-in brass cup plugs at the bottom of the main casting. This process was used through 1968.

main body to leak other than the factory pressed-in bottom plugs. During the machining process it was necessary to access several areas of the main casting from the bottom side. This mandated sealing them up afterwards. On the bottom rear portion of the main casting there are two large pressed-in plugs. Very early units have brass cup plugs.

The leakage rate for these units is almost 100 percent. Dabbing epoxy over them is similar to cutting your arm off and putting a Band-Aid on it! Over time, the glue comes loose and the leakage continues. A quick test to determine if the bottom plugs are leaking is to invert the casting and apply a soap/water solution to the plugs. Using compressed air and a long tip on the air gun, apply high-pressure air in through the main fuel-supply passages or pressed-in rear secondary disks.

If leakage is noted, the plugs are relatively easy to pull and re-install coated with epoxy or the casting tapped for custom-made screw in plugs.

The most reliable method to repair leaking bottom plugs: Custom screw-in plugs can be made from a piece of threaded rod or the end of a bolt. Slot the screw-in plug so it can be tightened with a screwdriver. With the threads coated, the epoxy is trapped in place. Simply dabbing epoxy over the leaking plugs does not work as a long-term repair. The epoxy eventually comes loose from the aluminum after many heating-up and cooling-down cycles of the engine.

Later units made after about 1968 used pressed-in aluminum plugs.

We still see a few of these that leak, but not nearly as often as the early-style brass plugs. The method to install and swage in the aluminum plugs continued to improve through the years of production. It is quite rare to find units made from about 1975 and later that leak.

The most reliable test for leaking bottom plugs: Apply a soap-and-water mixture over the plugs, then introduce high-pressure compressed air in through the front and rear jet openings in the fuel bowl. A small leak shows a tiny stream of bubbles. Large leaks as seen in this picture are easy to detect!

The swaging process was improved for later carburetors. It is rare to find a leaking bottom plug in the 1975 and newer units, but they should always be tested and repaired as needed.

In addition to pressed-in rear plugs, several pressed-in lead plugs are found under the main fuel-supply passages and one at a right angle to the accelerator pump checkball. It is rare for any of these plugs to leak, but on occasion they do and must be removed and the casting tapped for custom screw-in plugs coated with fuel-resistant epoxy.

Warped Air Horn/ Main Body

Many castings show warping across the front. This is most often caused by over-tightening the long front bolts. By design, the Quadrajet can withstand some warping across the front and still continue to work fine. Once the center of

Some air horns show warping across the front. The Quadrajet can withstand some warping and continue to work fine. In some cases the center of the air horn is raised up enough to cause fuel to seep around the center of the gasket. The air horn can be sanded or machined flat to re-conform to the main body.

the casting rises up and starts leaking, repair or replacement is usually required.

Occasionally, we also see some warping on the bottom of the main body and air horn. The Quadrajet uses a large, thick gasket between these parts. It is rare that the warping is severe enough to cause vacuum leaks or any other issues once the carburetor is placed in service.

Stripped Fuel Filter Housing Threads

This problem is quite common on early models, mostly because of the age of the material and how many times the fuel filter housing has been removed for filter replacement. There are several methods that can be used to repair this problem; the most effective is the installation of a Heli-Coil.

The Heli-Coil gives the filter housing hardened steel threads to tighten into; it is a very good repair. Several companies also market slightly oversized fittings with a self-tapping feature. They can be effective for minor thread-wear issues, but the Heli-Coil repair is more effective and permanent.

Once in a while a front plug shows leakage. Again, the best repair is to remove the old plugs and tap the casting for a screw-in plug. The plugs in the photo were made from hardened #8-32 screws, shortened so they do not extend too far into the casting and interfere with fuel flow from the main jets.

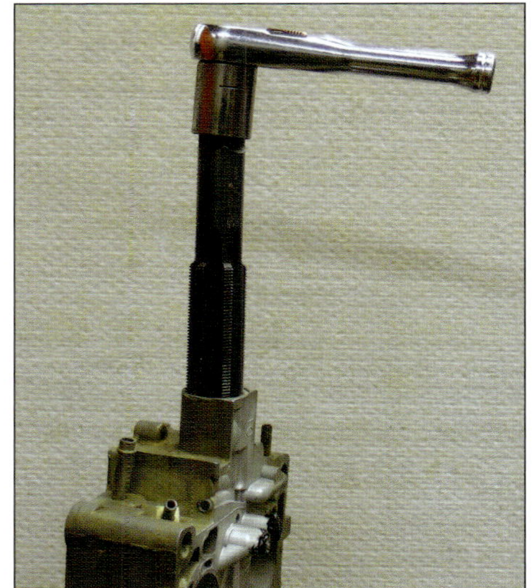

The best repair for stripped-out fuel-filter housing threads is to have a Heli-Coil installed. This is an excellent repair, and is much more reliable than installing aftermarket self-tapping fuel filter housings.

Loose Throttle Shafts

Almost all well-used units have some wear present at the primary throttle shaft. The factory did not install any bushings; the steel throttle shafts rest directly on the aluminum base plate. The steel wears out the aluminum even though later units used Teflon-coated shafts; they still wear out the base plates.

It is highly recommended that all units get new bronze primary shaft bushings. This insures that all incoming air goes through the carburetor. It helps with idle quality and tuning, as any air entering the engine at idle speed that comes in by the shaft is a vacuum leak. The air is also unfiltered. Unfiltered air can contain fine dust and abrasives that can be damaging to the engine. Installing bronze shaft bushings also reduces shaft side play. This ensures that the primary throttle plates always return to the same position when the linkage contacts the idle-stop screw.

A must for all rebuilt Quadrajets, installing primary shaft bushings reduces side play at the primary shaft. Leakage at the primary shaft is a big problem with used carburetors. It not only makes tuning nearly impossible, but it also allows unfiltered air to enter the engine. A custom self-guiding drill bit is used to open up the shaft bore in the base plate (first picture). A bronze bushing coated with Red Loctite is then driven in place (second picture).

Some companies offer bronze bushing-repair kits for the secondary throttle shaft. Since the secondary throttle plates seat tightly when closed, the carburetor can tolerate more play at the secondary shaft. It's quite common to see carburetors with a bit of side play at the secondary shaft. Provided the throttle plates are correctly indexed and centered in the bores when closed, installing bushings is not needed.

Taking the Carburetor Apart

Once you have selected a suitable core and it passes visual inspection, it's time to take it apart for cleaning, repair, and rebuilding. Close examination of the unit reveals a few minor problems in separating the three main parts if only the retaining screws were removed. Linkages run from one section to others and have to be removed prior to separating the components. The air horn should be removed first. The carburetor should be placed on a wooden spacer or other suitable fixture, so that any downward pressure applied during disassembly is not transmitted to the throttle shaft. The first step is to remove the link coming up from the base plate that moves the accelerator pump lever.

Early models use a clip to retain the arm, while later models require that the roll pin rotating the arm be driven back

Early carburetors have a clip on the end of the accelerator pump link. This makes for easy removal during rebuilding.

Later designs did not use a removable clip at the end of the link. The roll pin must be driven back to remove the link and arm. Drive the roll pin back just enough to remove the arm. This leaves enough room to get a small flat-tip screwdriver behind the roll pin. Pry it back through the arm when the carburetor is re-assembled.

Later-model carburetors used a #6-32 screw to hold the linkage to the choke shaft.

Early carburetors used a removable clip on the choke link. After removing the clip, the link can be turned 90 degrees and pulled up and out of the carburetor. This allows the air horn to be pulled straight up during removal.

against the housing. Very late units went to a larger link and back to a clip retaining it to the lever. Any linkage attached to the choke shaft must also be removed. Early models used a removable clip here as well; later models used a screw to retain the linkage.

It is advisable at this point to remove the secondary metering rod hanger and rods to keep from bending them when the air horn is removed. Also remove the threaded rod for the air-cleaner assembly. Once all the linkages are removed, remove all the screws that hold down

the air horn and gently pry up on the choke side if the air horn is stuck. Early models have a total of nine retaining screws, two long #10-32s in the rear, five shorter #10-32s, and two #8-32s inside the choke housing. Later units may have four additional #10-32 screws; make sure they are all removed prior to prying up on the air horn!

The air horn should be pulled straight up until the brass tubes clear the gasket. It may be necessary at this point to rotate the air horn to remove any linkage coming up from the choke pull-off if not previously removed. Once the air horn is clear of the main casting, lay it down on

Remove all of the screws holding the air horn to the main body.

choke pull-off mounted on the choke linkage; some early Buick carburetors also used a secondary choke pull-off on the same bracket.

Prior to removing the base plate, remove all items inside the main casting.

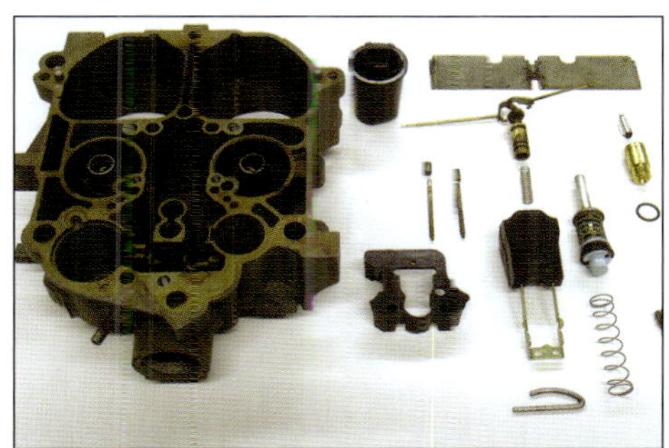

Once we get the carburetor stripped down, we are ready to go to work cleaning up the castings and small parts.

its top to avoid bending/damage to the brass tubes hanging from the bottom.

At this point it is best to remove any items bolted to the main casting. Some units have a choke mounted on the carburetor; others just have a choke linkage. If equipped with a hot-air choke or electric choke, the choke dial must be removed to access the choke-housing retaining screw.

Later models may use two choke pull-offs; one is mounted on the rear of the main casting, the other on the air horn. Most early units used a single

This includes the accelerator pump and spring, float splash-guard, float and fulcrum, power piston, primary metering rods, primary jets, AFT aneroid (if equipped), and the fuel-inlet seat. Some

A single screw holds the divorced-style choke assemblies to the main body. The choke assembly holds the choke pull-off and the fast-idle cam. The main shaft extends into the main body and engages with a lever that moves the link to the choke shaft. Don't forget to remove it from the main casting after removing the choke assembly.

Hot air and electric choke models have a single screw inside the choke housing to hold it to the main casting. These models had the choke pull-offs mounted on a separate bracket.

The correct removal procedure for the power piston is to depress and release it several times. The upward spring tension usually pushes the power piston and retainer out of the bore. If this procedure doesn't work, don't bend up on either arm of the power piston. Use a small screwdriver and pry up gently on the center of the arm as close to the piston as possible. Bent power-piston arms are one of the most common reasons that rebuilt Quadrajet carburetors do not work as intended!

carburetors contain other items, such as an auxiliary power piston and single metering rod. Don't forget to retrieve the power piston spring and the spring under the auxiliary power piston. Do not pry on either power piston arm to remove it. The correct method to remove it is to depress and quickly release the piston several times. If it becomes necessary to pry it from the casting, stay as close to the center as possible to avoid bending the arms that hold the metering rods.

If used, remove the APT-adjustment screw located directly in front of the power piston. Gently seat the screw, counting the number of turns up from seated. Record the number of turns and return the APT to this setting during carburetor assembly.

The checkball retainer and checkball can be removed at this time. If the checkball is stuck, a gentle tap with a soft-faced hammer or block of wood to the base plate will dislodge it. If not, a long scribe may be required to get it freed up, as it's common for corrosion to be present between the steel ball and aluminum casting if the carburetor has sat around for a long time or any water has found its way into the casting.

In some cases the main jets may be corroded into the casting and difficult to remove. Make sure to find a wide flat-tip screwdriver with a very tight fit in the jets, as they are easily damaged during removal. A slight tap on the screwdriver from a small hammer is often required to free them up in the casting. In extreme cases you may need to heat up the casting around the jets with a propane torch for about 30 seconds to help free them up.

Once the main body of the carburetor is stripped of all components, you are ready to pull the idle tubes. Although not mandatory for most rebuilds, we consider it a very good idea to remove them. They are most often corroded on the ends with lots of dirt and debris trapped under them.

Once all items are removed from the main casting, it can be inverted and

Prior to removing the APT screw, turn in clockwise until gently seated. Count the number of turns down from its original position. This will be our base setting when the carburetor is placed back in service.

The base plate uses either two or three screws to hold it to the main casting. Very late-model carburetors used a screw in the front of the base plate as shown.

Idle Tube Removal

Idle tubes are pressed into the main casting and hang into the main fuel supply passages after the main metering jets.

Factory idle tubes are a two-piece assembly. A collar is pressed over the tube and both parts are pressed into the main casting. The tiny restriction at the end of the idle tube is often plugged or obstructed with dirt and debris. A fair amount of foreign material also becomes trapped under the idle tubes. Failure to remove and clean in and under the tubes most often meets with poor results when the carburetor is placed back in service.

They are the first point of restriction in the idle system. The size of the opening in the bottom of the tubes is critical to correct carburetor operation. For applications with a larger-than-stock cam in the engine, removal and enlargement of the restriction in the tubes is required. The idle tubes use a two-piece design. The tube is made of brass: straight-walled .093 inch. At the bottom of the tube is a narrow section with a small hole. A larger brass collar is pressed over the tube and fits tightly in the carburetor main casting. The tubes are driven into the carburetor as a one-piece assembly. The idle tubes are long enough that the tips hang well into the fuel present in the main casting.

Removing the tubes is not difficult, but it does require some precision to ensure they are not damaged. The first step is to locate a small punch. It must be very close in diameter to the idle tube and slightly smaller than the inside diameter of the collar. Carefully locate the tip of the punch directly over the small tube inside the brass collar. With a few light taps from a small hammer, drive the tubes down. In most cases the tube moves down and the collar remains in place. Once in a while the collar tries to move down with the tube. If the tube moves down without the collar, drive it down only about ⅓ inch.

Next, you have to fabricate a tool to engage the collar so it can be pulled from the casting. Removing the tip from a drywall screw makes an effective tool to engage the brass collar.

A small, flat piece of metal should be placed over the top of the main casting to protect it during idle-tube removal. By using a small pair of wire cutters, we can clamp the screw and pry it up and out of the casting, effectively pulling the idle tubes with it. For stubborn idle tubes, heating the casting around the tubes just inside the fuel bowl for about 30 seconds with a propane torch helps removal.

Select or make a small punch with diameter of .090 to .093 inch. Very carefully drive the idle tube down just enough to engage the collar for removal.

Some idle tubes can be stubborn to remove. Heating up the casting right next to the idle tubes for about 30 seconds helps loosen them up. In most cases, they just about fall out after this procedure.

Idle Tube Removal CONTINUED

Once we have fully engaged the idle-tube collars with our modified self-tapping screw, gently pry up on the screw to pull the collar. Make sure to protect the top of the carburetor with a flat piece of metal.

In some cases the tubes and collar drive down together. Don't panic if this happens. Obtain a .093-inch drill bit and drill the tube down about ⅛ inch to facilitate engaging the screw and removing the tubes. Driving the idle tubes down more than ⅛ inch can cause damage to the small tips, as they may hit the bottom of the passages. If the tubes happen to come out of the collars and fall into the carburetor, the sharp tip of a drywall screw makes an effective puller to remove them.

If, for some reason, the idle tubes are not being removed, a small tool can be fabricated to clean them out. Find a small spring with a wire diameter of no more than .030 inch.

Few idle tubes are smaller than .030 inch, and the spring can be straightened out and a sharp tip put on it. This can be used to push through the idle tubes to make sure they are not obstructed. Plugged or partially restricted idle tubes are one of the single most common reasons that most Quadrajet rebuilds are not successful.

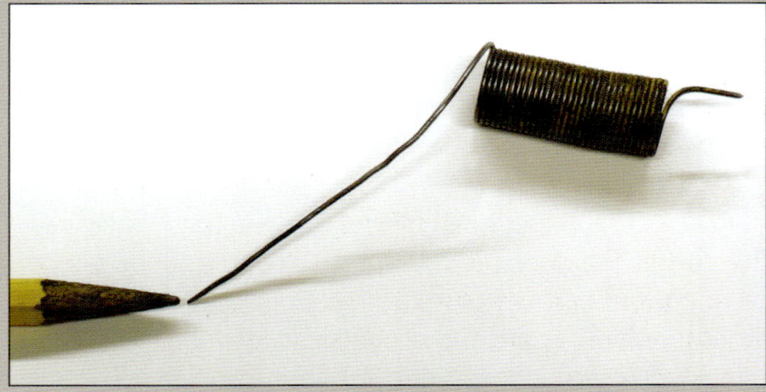

A quick cleaning tool for idle tubes can be made from a piece of spring wire. Carefully grinding a piece of .030-inch spring wire down on the tip makes a convenient tool for pushing out any debris trapped in the idle tubes without removing the tubes from the casting.

placed on our wooden block. There are either two or three screws retaining the base plate.

Remove the screws, but DO NOT pry up on the base plate. There are two guide pins formed from the main casting that are easily broken by uneven prying when removing the base plate. It is best to work from the rear portion of the base plate and tap or pry up gently on each side to keep it even. Too much on one side or the other almost always breaks one or both pins. This isn't a showstopper, but it does have to be addressed later. The pins control alignment between the main body and base plate.

Cleaning the Parts

Before you can begin assembly, all parts need to be cleaned. All the small parts can be cleaned by soaking in a small metal pan filled with carburetor cleaner. After soaking they should be blown off with high-pressure compressed air. A soft-bristle brush helps with really stubborn deposits and corrosion. The base plate, main casting, and air horn can be submerged for cleaning. Several companies offer solutions designed for this purpose. The chemicals used are mild and much more user-friendly than in years past. They also don't clean quite as well, but they still remove most of the heavy deposits. Once you have successfully cleaned the main components and all small parts, you are ready to put the carburetor back together.

It's a good idea to lay out all the parts after cleaning and do an inventory and close visual inspection. There are certainly a lot of small parts in the Quadrajet,

Submerging the carburetor and its components in a pail of carburetor cleaner for a few hours of soaking saves a lot of time in removing heavy deposits of grease and grime from the castings.

Several different styles of secondary linkage were used through the years of production. Many styles do not feature a positive stop to control the full open point of the secondaries, as seen on the top base plate. The bottom base plate has a boss that provides a positive stop for the linkage.

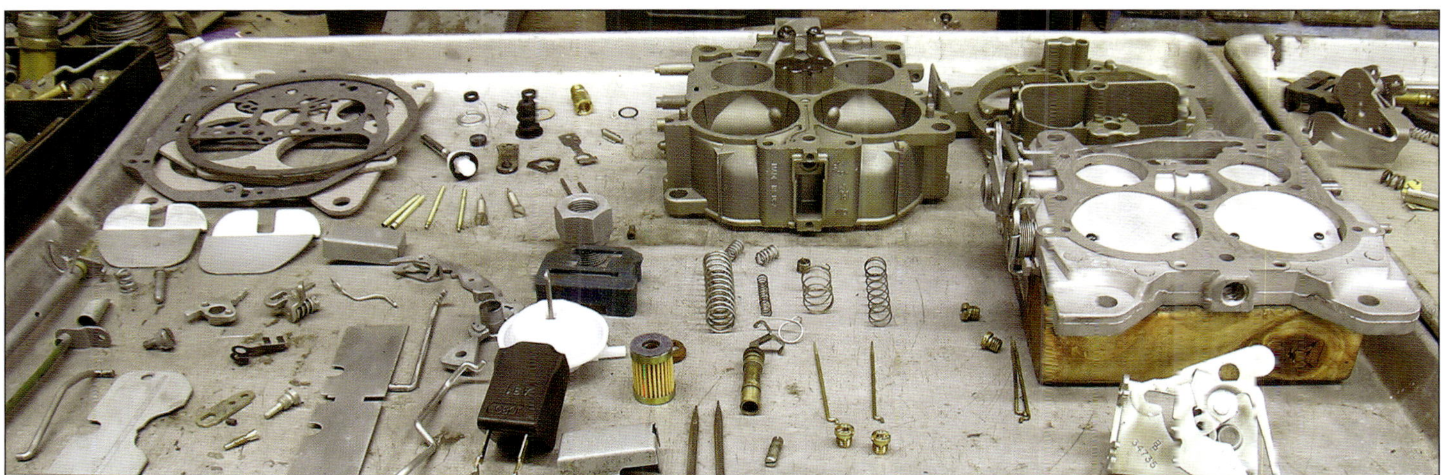

Our Quadrajet base-plate has been completely rebuilt. The main body and air horn have been plated. All small parts have been cleaned and plated as needed. We are now ready to start assembly.

and most of them do not come in rebuild kits and have to be re-used.

Base Plate

The base plate should be completely rebuilt first. Once attached to the main casting it provides a solid working platform when placed on a stand of small wooden blocks. Most throttle shafts have considerable side play in the base plate. It is essential to install shaft bushings to bring tolerance to a minimum. Excessive play allows unfiltered air to enter the engine, and it also acts like a vacuum leak. For your high-performance carburetor, you need to have full control of all of the incoming air for best results.

It is quite rare for any factory base plate to have either shaft correctly adjusted so that all four throttle-plates come to 90 degrees at full throttle. It's also common to find throttle plates that are not well-centered or indexed. Correct base plate rebuilding involves removing both shafts and all four throttle-plates. Clean and straighten the shafts as needed, then put them back in and adjust the linkage.

There are several different styles of base plates. Some provide a positive stop

Throttle Plate Stop

Many late-model carburetors used a base plate without a positive stop for the secondary throttle plates. You can make your own stop by grinding a "flat" on the base plate, then bending the linkage as shown until it stops the throttle plates at exactly 90 degrees.

We have modified the secondary lock-out pin to provide a positive stop when the throttle plates reach a full 90-degree angle.

for the secondary shaft, and others do not. For your high-performance carburetor, you need to provide a positive stop for the throttle plates at exactly 90 degrees.

Install the idle-mixture screws and gently seat them. If you have opened up the holes under the mixture screws we may find that the screws coil bind the springs under them well before the tapered ends seat into the base plate. It is necessary to clip a coil or so from the springs and continue to attempt to seat them until they do so before the springs coil bind. Failure to perform this step may cause tuning issues later, as you may not have full control of the idle fuel from rich to lean.

Attach the base plate to the main body. Make sure that you have chosen the correct gasket. Carefully line up every hole in the gasket with both parts. Some carburetor kits supply several different gaskets, as their kit may be designed to cover a broad range of carburetors.

Prior to installing any new parts, we need to check a few items. Two very common problems with rebuilt Quadra-jets are sticking power pistons and poor fuel-delivery from the accelerator pump. Both problems are caused by pitting, scratches, or corrosion on the bores in

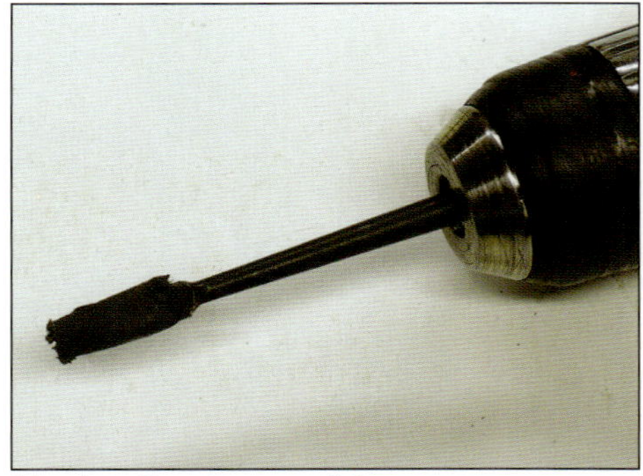

Honing the power piston removes any pitting or rough spots. This ensures that the power piston has free travel its entire distance up and down the bore.

the carburetor. The power piston must have free travel the entire length of the bore that it rides in. A quick way to polish the bore is to wrap a small section of 600-grit auto-body sandpaper around the tip of a rifle cleaning brush.

Test for fit into the bore and cut sections of the paper as needed. Once we establish a good fit, use a small variable-speed drill and polish the power piston bore, much like honing a cylinder in an engine. The fine sandpaper removes any high spots without damaging the bore. Use another piece of 600-grit paper to polish the piston. Test fit to make sure it has free travel the entire distance of the bore.

Perform a similar polishing operation for the accelerator pump bore if any pitting, scratches, or roughness to the surface are detected. A small brake hone with the stones wrapped with 600-grit paper makes a great honing tool for the accelerator pump bore.

Also check the two tiny holes leading to each pull-over supply well to make sure they are not obstructed. The range for these holes is usually between about

Opening up the holes under the idle mixture screws often requires that a coil or two be removed from the springs. This ensures that the idle-mixture screws can reach the fully seated position and provide full control of idle fuel.

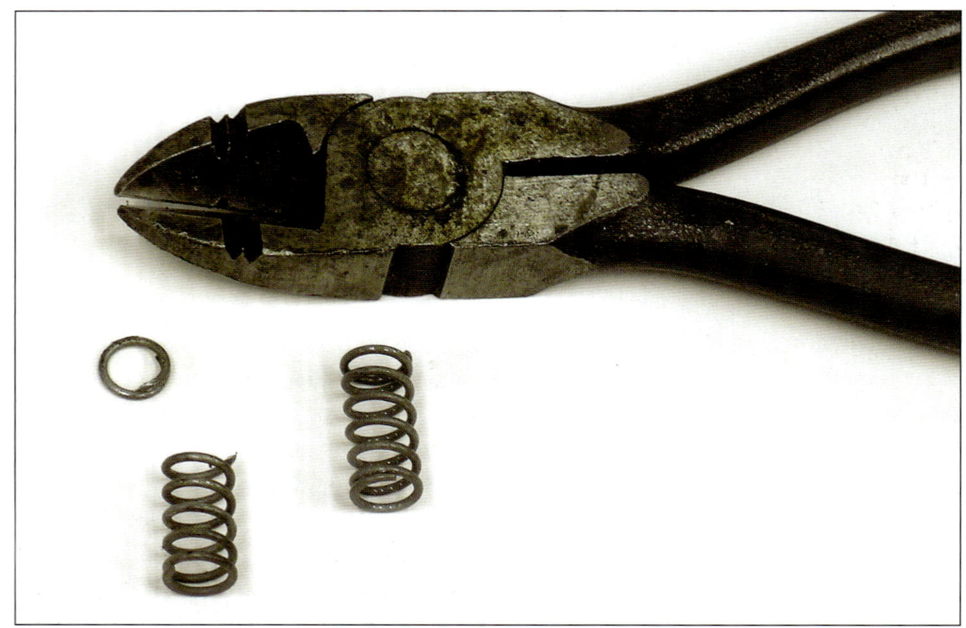

Using a piece of fine-grit autobody sandpaper wrapped around a small brake cylinder hone as a polishing tool for the accelerator pump works well. Removing any pitting or rough spots greatly increases accelerator-pump efficiency.

The tiny holes leading the secondary wells are often blocked or restricted with dirt and debris. Here we are cleaning and enlarging them to .040 inch with the drill bit held in a pin vise.

Alignment Pins

It is not uncommon for one or both of your dowel pins on the main body that line up the base plate to be broken. It is imperative that you have a good alignment between these parts. Insert two of the long #10-32 screws through the back two holes in the main body that attach them to the base plate. Also insert two ⁵⁄₁₆-inch bolts down through the large front holes in the main body. Check and adjust the alignment between the two parts. Then tighten the two or three screws that attach the two parts together.

On occasion, one or both of the alignment pins between the main body and base plate are broken. Installing the long 3-inch screws and using two long ⁵⁄₁₆-inch bolts keeps both parts lined up correctly while the base-plate screws are tightened.

.033 inch and .040 inch. Any size drill bit in that range can be used to clean out the holes, but going any larger than .040 inch is not needed for any application.

It's also a good idea to carefully chase the threaded holes in the main casting with a tap, especially if the main casting has been bead blasted during the cleaning process and/or re-plated. Use the appro-priate #10-32 for all but the two holes next to the primary bores – they require a #8-32 tap. Blow out the holes with compressed air; make sure to wear safety glasses!

Once we have successfully attached the base plate to the main body and have good alignment between the two parts, you are ready to start installing parts into the main body. Install the choke and choke hardware first. Using a block of wood as a support, lay the carburetor on its left side. Shim as needed so the carburetor is lying flat on the block. With a long pair of needle-nose pliers, guide the choke lever inside the casting so that the squared end lines up with the hole in the casting. A #2 Phillips screwdriver secures the lever and turns it to facilitate installation of the choke housing.

A #2 Phillips screwdriver fits perfectly in the choke link to line it up prior to installing the choke housing.

Hot-air-style choke housings used a small plastic sleeve between the housing and the main casting. The sleeve seals off the vacuum passage between the two parts and helps to align them during assembly. This passage should be blocked off if the carburetor uses or is being converted to an electric choke.

Prior to installing the choke housing or divorced-choke hardware, open the primary throttle shaft off the idle position so the fast-idle cam doesn't interfere with the adjustment linkage. For hot-air choke models, make sure the small plastic tube is installed in the vacuum passage.

The tube lines up the choke housing and acts as a seal to prevent vacuum leaks. Once in place, tighten the choke housing or bracket, holding the pull-off and fast idle cam (divorced choke models) with the appropriate #10-32 screw. Move the choke lever and check fast-idle cam operation. The fast-idle cam should line up and engage with the lever on the primary throttle shaft.

Install the main jets first, using a screwdriver that has a good fit to avoid slipping and damaging them. Make sure they are fully seated. It may be necessary to chase the threads in the main body with a ¼ -28 inch tap.

Using an old checkball, carefully drop it into the passage below the checkball retainer.

Give it a couple of taps with a long punch and small hammer; invert the carburetor to remove the checkball. It may require a tap on the bottom of the base plate to dislodge the checkball. We have just formed a good seat for the new check-ball. Insert the new checkball and retainer. This step should never be left out, as a leaking checkball/seat can cause the engine to hesitate as the accelerator pump has to fill the passages above the checkball before it delivers a good shot of fuel to the engine once the carburetor is placed in service.

Install the fuel-inlet assembly next. Make sure to use a new seal and that the old seal was removed. It is pretty common for the old seal to stick tightly, and it may not be readily visible if the black coating is no longer present. A scribe makes a good tool to remove the old seal. Also closely inspect the threads; many early units may require a Heli-Coil here as the threads are often eroded away or stripped out. Tighten the fuel inlet with a well-fitting screwdriver. Make sure the needle and seat have been previously vacuum tested. The rejection rate for new needle/seat assemblies is about 1 in 10. Do not install any other type or style of fuel-inlet needle/seat assembly. Some low-end kits come with a combination valve to replace the original components. They are extremely restrictive as well as unreliable in long-term use.

If you are working with an early carburetor that originally came with the old plunger-style fuel valve, replace it with the

Drop an old checkball into the accelerator pump checkball well and give it a quick tap with a punch. This forms a new seat for the checkball and keeps the accelerator passages full of fuel for maximum fuel delivery from the accelerator pump. Turn the casting over and remove and discard the old checkball. A light tap from a hammer may be required to dislodge the checkball from the new seat.

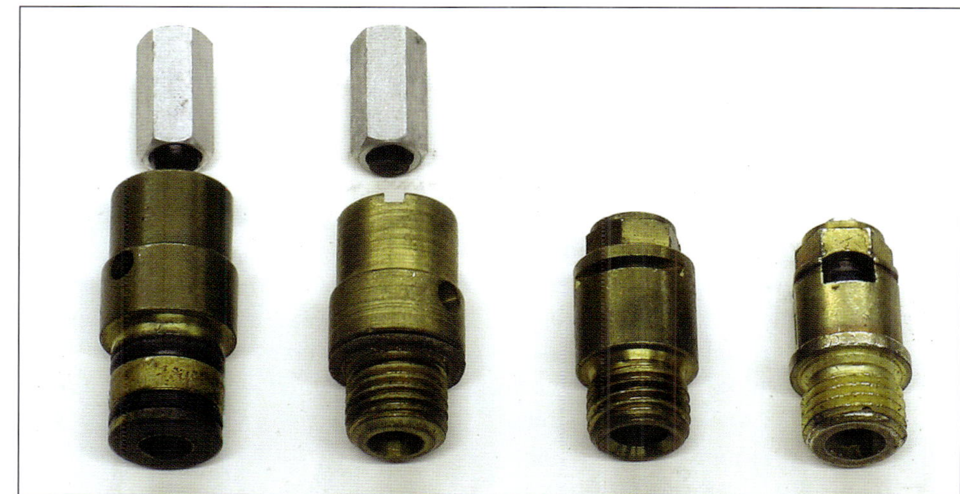

Some rebuild kits may contain aftermarket replacement needle/seat assemblies. These are very restrictive and should never be used for high performance.

later-style components. We do not use or recommend the early-style carburetors for serious high-performance work due to several shortcomings. The fuel-inlet design and float-fulcrum arrangement top the list. We're not recommending that one avoid all of the early carburetors completely, but a later unit is a better choice and typically much less labor-intensive to build, with a better end result.

Install the float and fulcrum and check the float height. The float height should be set to stock specifications for stock rebuilds. If the stock settings cannot be found, set the float to exactly ¼ inch for all models. The tiny clip on the

end of the needle can be discarded (it is not needed). Also, check to make sure the top portion of the fulcrum is slightly higher than the top of the main body. This ensures that it is firmly held in place by the air horn after assembly.

The power piston and metering rods are next. If removed, re-install the APT adjustment screw on late-model carburetors and set to the manufacturer's recommended setting.

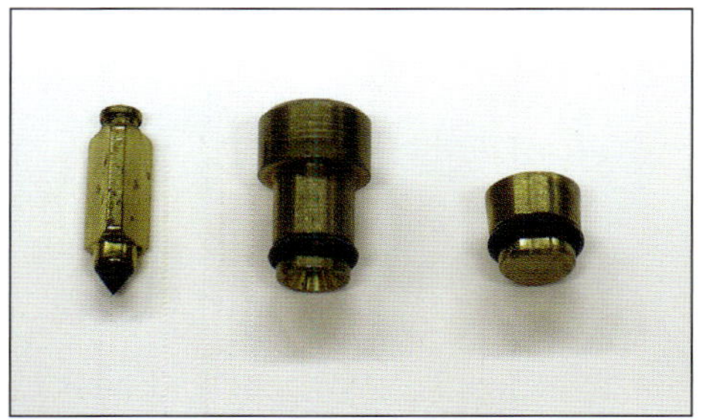

Kits are available to upgrade the early-style plunger fuel-valve assemblies to the later fuel-inlet-seat-and-needle design. The kit shown is from the Carburetor Shop in Eldon, Missouri.

Install the APT adjustment screw into the main casting. Make sure to put the spring in first and return the APT screw to the original setting recorded when the carburetor was taken apart.

Make sure that the metering rods hang exactly even for all models. Most of the hanger arms have been bent from incorrect removal during prior rebuilding. Place the power piston spring in the bore first, then very carefully insert the power piston and guide the metering rods into the jets. The splash-guard has to be installed first for later-model carburetors, and this can make this step a bit more difficult. The plastic retainer should be carefully pushed into the top of the bore while holding the power piston in the down position. It may be necessary to gently stake the top of the bore to keep the retainer in place.

Very carefully stake the top of the power piston bore to hold the retainer in place. Test the power piston for free travel afterward.

Once in place, check for free movement of the power piston the full length of the bore. It should move freely through the entire distance. If not, remove the piston and check for scratches on the polished surface. This identifies any defects in the bore. Hone the power-piston bore again and polish the piston and repeat installation until the piston moves freely.

Install the splash-guard over the float assembly. On some early-model carburetors the splash-guard can be considerably warped and may interfere with the float arms. This is because the carburetor fills with fuel and the float rises to the top of the fuel bowl. Check for sufficient clearance.

Power Piston Retainer

If considerable damage is present at the top of the power-piston bore and it no longer holds the retainer, we have the option of using a long, thin piece of metal (a machinist's ruler works well) to hold the piston down while the air horn is installed. You can then slide out the piece of metal while holding the air horn down against the gasket.

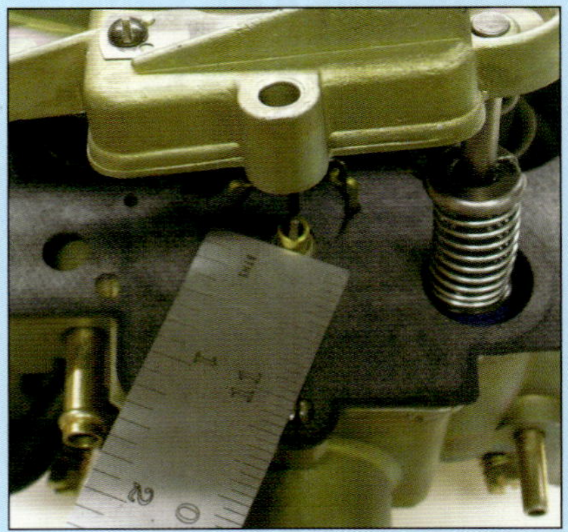

Some carburetors have damage to the material around the power piston retainer. In lieu of staking, we can hold the retainer in place with a machinist's ruler during carburetor assembly.

Install the top gasket next. The center portion of the gasket is split to allow it to be slid under the power piston arms without bending them. Some gaskets have a small hole where the accelerator pump shaft goes through them. Guide the accelerator pump shaft through the gasket and make sure that the pump seal stays in place over the return spring. Guide the rear outboard portions of the gasket onto the pins on the main body. You are now ready to install the air horn.

Air Horn

The secondary air-flap shaft should be checked for free movement through its range of motion. In the center of the shaft is a small plastic cam that raises the secondary metering rods. It should have a tight fit on the shaft. Some cams may be worn and/or cracked.

Replacing the cam involves removing the staked portion of the screws that hold the air flaps in place and unhooking the spring from the pin on the end of the shaft. The shaft is turned to line up the pin with the slot in the casting, and the shaft is slid out to gain access to the cam. It's not uncommon to find the spring is rusted and/or has lost its tension. This is a trouble area for Quadrajets that is most often overlooked. A worn spring causes secondary operation issues once the carburetor is placed in service. Secondary cams and springs are not part of any rebuild kits, but they are still available as special-order items through various suppliers.

Reinstall the secondary air-flap shaft. Line up the air flaps, move the shaft side to side to find the center of travel, and secure the flaps with new #6-32 screws. Use a drop of Loctite and stake the exposed portion of the screws. The air flaps should move smoothly through their entire range of motion. If not, repeat the centering and indexing procedure until they move without any binding or tight spots.

Using a .026- or .028-inch drill bit and pin vise, very carefully clean out the accelerator pump discharge holes in the

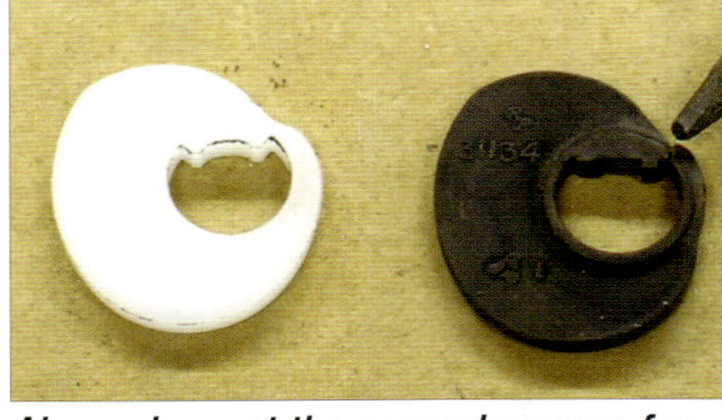

Always inspect the secondary cam for cracks and wear. It is best to always install a new cam. They are not available in most rebuild kits.

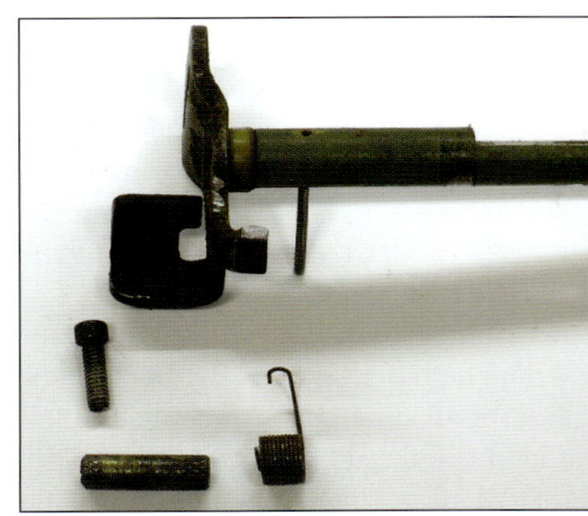

The secondary air flaps are controlled by an adjustable spring. The adjustment screw is held in place by an Allen-head set screw coming in from the bottom of the casting.

air horn. Some applications may require larger holes for a quicker pump shot; this is covered in the next chapter.

Check the choke flap and shaft (if being used) for free movement over its

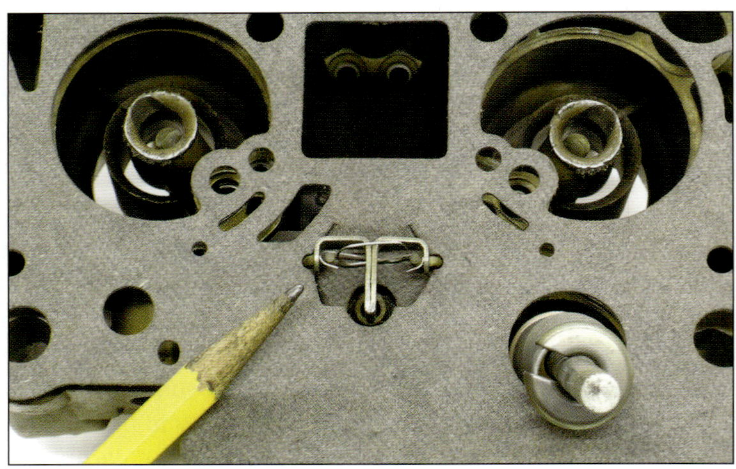

range of motion. Make sure that the flap is completely sealing off the housing when closed and not sticking. If any binding or tight spots are present, carefully loosen the two attaching screws about ¼ turn and adjust as needed.

The brass tubes in the air horn need to be reinstalled if removed earlier. Most models have four tubes; some later-model carbs that used the primary pull-over enrichment system have six tubes.

They need to be gently driven into place. A drop of Loctite and staking prevents them from falling out.

The air horn is now ready to install. It is recommended that any linkages attaching the choke pull-off, choke, or accelerator pump be installed after the air horn is installed and tightened down. For some models you have to install the link from the choke pull-off while the air horn is held at a right angle to the main body, and then carefully lowered into position.

Make sure that the brass tubes go through the appropriate holes in the air horn gasket and that the accelerator pump shaft goes up through the correct hole in the air horn. It often takes a bit of patience to install the air horn to get everything lined up. The goal is to install it without damaging any parts. If the power piston retainer happens to pop out of the bore, installing the air horn could bend the ends of the metering rods over as they come out of the jets. Never force the air horn into place. If it doesn't seat easily on the main body, something is probably wrong; correct the problem before continuing.

Install the two #8-32 screws inside the choke housing first and gently seat them. The air horn should be down tightly against the main body. Before installing the rest of the retaining screws, push down on the accelerator pump to make sure it has full travel and returns easily to the up position. Install the remaining attaching screws; tighten the front screw, then the two behind the choke housing. Tighten the rear out-

Some late-model carburetors used Primary Pullover Enrichment (POE). They are easily identified by the two extra tubes in the air horn.

board and long screws in the back of the carburetor last.

Before attaching any linkages, install the fuel-filter housing and gasket. Invert the carburetor and vacuum-test the needle-seat assembly. If it does not hold vacuum when inverted, then it will not hold back fuel when placed in service. Make sure to vacuum-test the needle-seat assembly before installing a fuel filter with a built-in check valve.

You can now reattach any choke linkages and the accelerator pump linkage. If removed, reinstall the accelerator pump arm. Use a small punch as a guide while prying the roll pin back through the arm.

Bolt on any other items such as the secondary choke pull-off and idle stop

All needle/seat assemblies should be vacuum tested prior to being placed in service. At least one in ten do not pass a vacuum test.

Most early choke pull-offs required bending the linkage to adjust the angle of the choke flap. Although somewhat primitive, they work quite well once the correct settings are found.

Later designs used an adjustment screw on the choke pull-off to fine-tune the choke-flap open angle on initial engine start-up.

solenoid if used. Adjust the choke pull-off to the manufacturer's recommended specifications. This information is usually provided in the instructions with the rebuild kit.

Early carburetors may require bending the choke link and/or the arm coming up from the choke linkage. Later carburetors used adjustable choke pull-offs.

Rebuild Kits

A wide variety of rebuild kits are currently available for Quadrajet carburetors. The vast majority of these kits only contain gaskets and a few small parts, such as clips, seals, and a checkball. Most contain a new seal for the old accelerator pump instead of a new pump with a seal. They may also include an aftermarket replacement for the OEM-style fuel inlet seat and needle assembly. Most of these assemblies are very restrictive and do not supply sufficient fuel to the carburetor for any sort of high-performance application. Always use the factory-style fuel-inlet and needle assembly. In order to make rebuild kits more universal, many companies include several different gasket sets. This allows a single kit to cover a

To make rebuild kits more "universal," they often contain an assortment of gaskets. Make sure to use the correct gaskets for your application.

broader range of applications. Always match up the gaskets in the kit to the ones removed, and check alignment of all holes between all of the components.

A new float is usually not included in a rebuild kit, and must be purchased separately. Factory-style floats should always be replaced during carburetor rebuilding. Modern fuels can be incredibly hard on the factory floats. They may have soaked up some fuel and be "heavy." A "heavy" float does not provide accurate fuel control and is the single

Modern fuel requires the use of carburetor parts that are compatible with ethanol. These custom rebuild kits from Cliff's High Performance in Mount Vernon, Ohio, are not only the most complete and correct kits in this industry but they also contain components that have a lifetime warranty with any fuel. In addition, Cliff's High Performance (cliffshp@embarqmail.com) offers a selection of difficult-to-find parts and a broad selection of high-performance and tuning parts for Quadrajet carburetors.

most common reason that rebuilds meet with negative results.

Choke pull-offs are also sold separately. Always vacuum-test the factory choke pull-off and replace as needed. It may be necessary to adjust the new choke pull-off for correct choke flap unloading on cold start-ups.

Several companies offer custom rebuilding kits. These kits include additional components such as a power piston spring, plastic secondary cam, secondary air-flap spring, and a new accelerator pump. Most custom kits are application-specific and contain the correct OEM-style components for the carburetor number we are working with. They are well worth the additional expense.

Always vacuum test the choke pull-off and replace as necessary.

HIGH-PERFORMANCE MODIFICATIONS

Now that you've learned how the Quadrajet works, how to correctly rebuild it, and what different models were available, it's time to apply some high-performance modifications. With the correct modifications in the right places, the Quadrajet has no trouble being up to the task of powering your new engine combination. Even if you are still working with a basically "stock" engine, the modifications mentioned in this chapter help improve performance, driveability, and fuel economy.

Base Plate Modifications

The first place to start is the base plate. It is quite rare to find a base plate where the throttle plates reach a full 90-degree angle at full throttle. It's also quite common to find units with bent shafts and throttle plates that don't fully seat in the closed position.

Once you've cleaned the base plate and installed bronze shaft bushings, hold it up to a light to insure that all the throttle plates are fully closed. This should be done prior to putting Loctite on the screws and staking them so they don't fall out. Once you're content that the throttle plates are tightly shut and the shafts are indexed, open the linkage to full open and check to see if the throttle plates all reach a full-open position. A machinist's ruler is helpful here, but we can do a pretty good job by simply observing them. The eye picks up the difference when there is not an even distance from both sides down to the base plate flat surface.

This custom high-performance Quadrajet has been tested and is ready for shipping. Note the modified fuel-filter housing with the -8AN fitting.

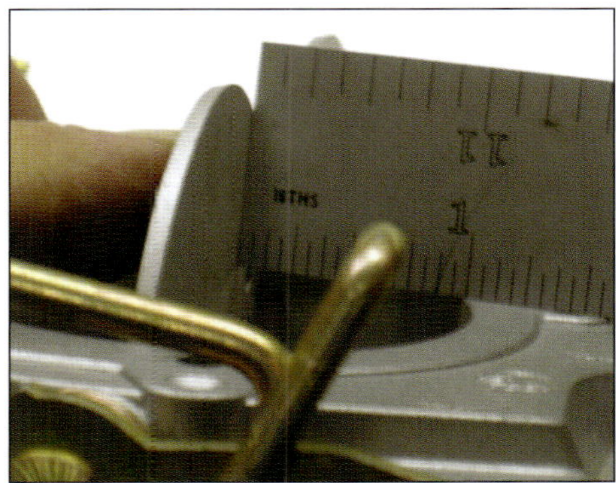

It's quite rare to find a factory carburetor with all the throttle plates opening to a full 90-degree angle. A machinist's ruler laid flat on the bottom of the casting makes a great tool to determine when the throttle plates are fully open.

Most primary throttle shafts rotate far enough to have the throttle plates well past center at full throttle. A pair of pliers is used here to bend the throttle stop for a full 90-degree opening.

The first thing to do is to create a stop for the secondary shaft if one is not present. Many early units had a stop as part of the base plate and linkage, but later units did not. The only way to guarantee that the secondary throttle plates always reach a full-open angle is to provide a positive stop for them. Most models that do not have a positive stop do have secondary lock-out linkage on the right side of the secondary shaft. Grinding a small, flat area followed by bending the short side of the lock-out rod creates a positive stop to ensure that the throttle plates always reach a full-open position.

Next, make sure the primary throttle plates open fully. Most roll past center. Bending the linkage where it is contacted by the throttle speed screw acts as a stop when the linkage hits the base plate at full throttle.

After this procedure you typically find that the linkage is not opening the secondaries all the way. There were several styles of linkage used to apply the secondaries. Observe the contact between the parts and gently bend as needed to get the secondaries to a full-open position just as the primary throttle shaft linkage contacts the base plate. This procedure often takes a few minutes, but it's well worth the effort. We've observed an engine power loss during dyno testing any time the throttle plates did not reach a full-open position.

Once you're happy with the throttle angles at full-open and the throttle plates are fully closed at rest, you can move on to other base-plate modifications. Remove the idle-mixture adjustment screws and measure the size of the hole under them. Most late-model carburetors have very small holes present here. A good starting point is to drill the holes to .090 inch. This is fine for most applications and makes sure that it does not require turning the mixture screws out four, five, or even six turns to get enough idle fuel to the engine. Having .090-inch holes under the mixture screws is not always

A positive stop is added to ensure that the throttle plates always open and stop at exactly 90 degrees.

Once the primary throttle plates are adjusted for a full open angle and a positive stop for the secondaries is made, bend the linkage as needed to make sure the secondaries reach the full open position.

Slotting Idle Mixture Screws

Many late-model Quadrajets used idle mixture screws that require special tools to turn them. In lieu of buying or making special tools, the heads of the mixture screws can be quickly modified so that they can be turned with a small flat-tip screwdriver.

This simply involves gently clamping an idle-mixture screw in a small soft-jawed vise and slotting the head with a fine hacksaw blade. Idle mixture screws are often case hardened. It helps to gently remove a few thousandths of an inch of material from the surface before attempting to slot them with a hacksaw blade. This can be done with a bench grinder or a cutting wheel on a high-speed air grinder. Once you have removed a small amount of surface material, they are much easier to modify. Keep the cut with the hacksaw blade in the center of the screws and only go deep enough so that they can be easily turned with a small screwdriver.

Adding idle bypass air is seldom needed with near-stock engines and very mild camshafts. The amount of bypass air to use varies greatly for different applications. For near-stock engines with no compression-ratio increase and a slightly-larger-than stock cam, start out with .050 to .080 inch of idle bypass air. It is recommended that the holes be drilled slightly smaller than the desired size with a power drill, then final-sized to the desired dimension by hand using a pin vise to hold the bit. Also make sure that holes are drilled in the main body in the passages leading down to the bypass air holes in the base plate. Most early carburetors do not have the main body drilled or have provisions in the base plate for bypass air. It can be added, but precision measurements and drilling is required. For the early carburetors, it's often just as easy to drill small holes in the primary throttle plates if bypass air is required. Although not quite as effective as using the bypass air system, it does help to keep the primary throttle plate angle lower at idle with big cams. Many late-model carburetors have the idle bypass air system in place but no holes drilled through the passages in the base plate.

Once you have completed the needed base-plate modifications, it should be fitted to the main casting. Perform a quick test-fit without using a gasket to make sure that the epoxy you have used to seal the bottom plugs is not holding the base plate so high that the gasket does not seal. Any epoxy on the sides of the plugs may also affect alignment and not allow the guide pins on the main body to locate the base plate.

Once you have a good test fit, check the gasket for alignment on both the main body and base plate. There were quite a few different gaskets used between the main body and base plate. Some rebuild kits come with several styles; make sure to match up each hole between both parts.

Always test fit the base plate to the fuel bowl after applying epoxy to the bottom plugs. Too much epoxy prevents the gasket from sealing between the parts and causes a vacuum leak.

needed, especially on stock rebuilding of late-model emission carburetors, but it doesn't hurt anything and often provides more control of the mixtures at a much lesser setting once the carburetor is placed in service. After drilling the holes, gently seat the mixture screws. In most cases the springs under the screws coil bind before the screws seat. Clip about one coil from the springs and seat them again. Continue clipping small amounts from the springs until they are fully seated. Failure to do so causes loss of control of the idle mixture once the carburetor is placed in service.

Idle bypass air was used on many models. If idle bypass air was not used, most carburetors have all of the passages present.

Many late-model carburetors do not use idle bypass air even though all the passages for it are in place.

How to Install Idle Bypass Air

Many very early production carburetors do not have a bypass air system. It can be added at some additional effort. There are basically two methods. The first is to locate the voids in the main body that lead down to the base plate. If holes are not present to allow bypass air, they must be added. This requires drilling them to at least .125 inch, but they can be up to about .200 inch. Next, you must obtain a gasket with the holes present that allow air to travel from the bottom of the voids to the base plate. The gasket used acts as a guide for the next set of holes. A drill press is needed here to drill the holes in the base plate.

Adding idle bypass air to an early-style base plate requires exact precision. The drill press should be set so the bit does not go all the way through the casting.

They should be drilled almost all the way through the base plate, but leave at least .100 inch of material. The next step is the most difficult. Small holes must be drilled at an angle from under the throttle plates to intersect the two holes you just provided in the base plate. This allows air coming down from the main casting to route under the throttle plates at idle.

An alternative method is to drill your holes clear through the base plate and grind channels that lead under the throttle plates. Either method works equally well. The last step is to make sure that the air horn has two small channels in it to provide air a path to reach the voids just above the air horn gasket.

Small holes are drilled from just under the throttle plates to intersect the holes drilled in the previous photo. This allows air to enter the engine at idle speed below the throttle plates, and improves idle quality for engines with radical camshafts.

When adding bypass air, make sure two channels in the air horn allow for some air to enter the voids leading in the fuel bowl that lead down to the base plate.

Having selected the correct gasket, install the gasket onto the bottom of the main body. Carefully push the gasket over the alignment pins, and then attach the base plate with two or three #10-32 screws as needed. Check for full opening of the throttle plates. You are now ready to make main-casting modifications.

Set the assembly on a small block of wood. Make sure that the linkage is not resting on the workbench, as we will be making several light blows to the casting and the linkage is easily damaged.

The first modification is quite simple. Obtain a used, steel checkball in good

Several different base plate gaskets were used. Some rebuild kits contain two or more gaskets. Always test fit and check all passages between the fuel bowl and base plate to make sure the correct gasket is used.

After forming a new checkball seat, install a new checkball and fill the passage above it with solvent or fuel. Any leakage at the checkball seat causes a delay in fuel delivery from the accelerator pump.

works fine), fill the cavity above the checkball. It must hold the fluid without leaking down.

If it does not, the passages after the checkball leading to the openings in the air horn will not stay full with fuel. This can cause a slight stumble or hesitation from the engine when the carburetor is placed in service.

Carburetors that have been exposed to a lot of water and are heavily oxidized may require several attempts to form a new checkball seat. This is a very important step in carburetor preparation. Continue the process until the ball effectively seals, or obtain another casting.

Once the checkball seat modifications are successful, install the checkball retainer and seat it tightly into the casting. Apply a soap/water mix to the sealing area and apply some compressed air into one of the accelerator pump discharge passages at the top of the casting. Any bubbles at the retainer indicate a leak. It may be necessary to seat the retainer again or apply a small amount of Marine Tex to the threads to form a perfect seal. Any leakage at the retainer negatively affects carburetor performance.

The main jets are next on the list. Carefully chase the threads for the jets

A ¼-inch -28 tap is used to clean the threads in the fuel bowl prior to installing the jets. Any defects on the threads can damage the threads on the jets and prevent them from fully seating.

condition. Drop it into the checkball location in the main body. Using a small flat punch and a hammer, seat the checkball. Three or four light taps is usually sufficient. Invert the casting and gently tap the base plate to dislodge the checkball. Turn the casting back over. Using light solvent (carburetor cleaner

The two small holes in the fuel bowl leading to the secondary accelerator wells are easily plugged with dirt and debris. Always clean these holes out with a .035- or .040-inch drill bit.

Most late-model carburetors have small holes in the base plate under the idle-mixture screws. This limits idle fuel capabilities even after the caps over the idle mixture screws were removed to provide access for idle-fuel adjustments.

with a ¼ inch -28 tap. Clean the area where the jets seat in the casting. Gently thread in the jets and seat with a well-fitting screwdriver.

Next on the list is to clean/open up the feed-holes to the main accelerating wells. The holes are typically between .035 and .040 inch. We've seen them drilled as large as .250 inch in attempts to increase carburetor-bowl capacity. It is not recommended to go past .040 inch for any application, as this system can continue to feed fuel at full throttle and affect metering.

Modifying the Idle System

Modifying the idle system is the most important part of setting up a high-performance Quadrajet carburetor. It does not matter how well the carburetor runs at heavy and full throttle if it doesn't have the engine idling well and providing clean/crisp off-idle performance. The vast majority of Quadrajets we encounter as cores have very limited idle-fuel capabilities. Typically, the newer the production date of the carburetor, the more stringent the emission standards were, and the more restrictive the idle-fuel capabilities are.

Before you make any modifications to the idle system, you have to look closely at the parameters of the engine that it is to be used on. The most important factors to consider are cubic-inch displacement (CID), the static compression ratio, and the camshaft specifications. These three items dictate the engine characteristics at low engine speeds and the amount of idle fuel and bypass air that is required from your carburetor for positive results.

In this section we talk about six basic items: the idle tubes, idle-air bleed(s), idle down-channel restrictions, idle bypass air, and holes under the mixture screws. If you are not familiar with these items, go back and study Chapter 2. As part of the rebuilding process, the idle tubes should have been removed from the main casting. If not, study the section in Chapter 5 that describes idle-tube removal.

Prior to modifying the idle tubes, you need to look at the size of the idle-air bleeds. Quadrajets use one pair of idle-air bleeds in either the upper main body just above the idle channel restrictions or in the air horn right above the channel restrictions. Either location works equally well, but any one carburetor cannot have both upper idle-air bleeds. Neither location really has any advantage over the other, but the sizes used vary some to get the same end results. There is a second

These idle tubes have the collars driven back into place and are ready for re-sizing and installation back into the carburetor.

pair of lower idle air bleeds in the main body in line with the fuel passage leading down from the idle channel restrictions.

Before making any modifications, size both the upper and lower idle-air bleeds. In most cases the lower air bleeds are between .070 and .080 inch. This is fine and it is seldom (if ever) necessary to modify them. The upper air bleed just above the idle channel restriction is most often between .060 and .070 inch. We've seen some samples as large as .090 inch, mostly on early-'70s Chevrolet carburetors.

It is now time to measure all of the idle-system components and evaluate your application in order to set up the idle system to work correctly. Measure and write down all of the items in the idle system in the following order:

Idle tube:
Idle down channel:
Upper idle air bleed:
Lower idle air bleed:
Idle bypass air:
Holes under idle mixture screw(s):
Holes in throttle plates (if used):

For reference we've inserted the following numbers from our carburetor.

Idle tube:	.033 inch
Idle down channel:	.040 inch
Upper idle air bleed:	.070 inch
Lower idle air bleed:	.070 inch
Idle bypass air:	NONE
Mixture screw holes:	.070 inch
Holes in plates:	NONE

Now you need to look closely at your particular application and make the necessary modifications to all of the areas listed above. For reference we start out with a very "mild" application— 350ci engine, 9:1 static compression ratio, and a 204/214/112ICL camshaft. This is a very common application using a small engine, conservative static compression ratio, and "RV"-type camshaft. The 350 CID engine with a compression ratio of 9:1 and a 204/214 @ .050 inch duration cam has very similar characteristics to a "stock" 8:1 compression-ratio engine using the factory-type cam.

Most late-model 350ci engines used cams in the 195 to 205 @ .050-inch-range with a static compression ratio of about 8.2:1. Going up to 9:1 compression ratio and adding about 10 degrees more cam has the new combination exhibiting very similar idle and off-idle qualities as the original design, with improved power at all RPM and a slightly extended RPM

range. For this particular application the stock carburetor settings are very close. We can still help it out slightly. For starters, let's use the following combination:

Idle tube:	.036 inch
Idle down channel:	.046 inch
Upper idle air bleed:	.070 inch
(.050 inch if located in the air horn)	
Lower idle air bleed:	.070 inch
Idle bypass air:	.080 inch
Mixture screw holes:	.090 inch
Holes in plates:	NONE

The only modifications we made for this combination were to increase idle-tube size by .003 inch, idle-down channel by .006 inch, mixture-screw holes by .020 inch and add some bypass air. If the carburetor we are using has a hot-air-style choke and it is being retained, it would be best to not add the bypass air. This is because the vacuum supplied to the choke housing does the same thing. If we have eliminated the hot-air-style choke and or converted it to electric and blocked off the vacuum supply passage, then the bypass air will be beneficial.

For our second example, let's take the same engine and add a bigger "high-performance" camshaft. For starters, we leave the static compression ratio at 9:1 and install a 214/224/112 camshaft. This much larger camshaft now has our 350ci engine operating at a lower vacuum signal at idle and low speeds. To compensate, we increase the size of a few of the items in our idle system accordingly. This adds the needed fuel for good idle qualities.

Idle tube:	.037 inch
Idle down channel:	.052 to .055 inch
Upper idle air bleed:	.070 inch
(.050 inch if located in the air horn)	
Lower idle air bleed:	.070 inch
Idle bypass air:	.080 to .085 inch
Mixture screw holes:	.096 inch
Holes in plates:	NONE

As seen, the increase wasn't all that dramatic, but it is beneficial for the new application. We provided a range of sizes for the idle-down channel and the idle by-pass air. It's best to start out with the smaller sizes, as we can always go back in later and enlarge them. If the static compression ratio were increased about one full point when the cam was changed to 214/224 @ .050 inch, the vacuum loss at idle and low speeds would have been

Use the small drill bits to check the size of the main air bleeds. Most Quadrajets used two pair of main air bleeds. One pair is in the main casting (shown here), and another pair in the airhorn.

negligible. You could have used the first example as a starting point, and then increased the sizes if needed.

The most important thing to keep in mind when coming up with a "recipe" for your carburetor is that very small changes can have a significant effect on the results. In addition, almost every application is somewhat different. In this hobby it is very difficult to build any two engines to exactly the same specifications, or duplicate exactly what someone else is doing. You might even get all the specifications and correct parts to build your engine, and still end up with slightly different results. Little things such as deck height, camshaft timing, and torque-converter efficiency affect the engine's operating characteristics when it is placed in service.

Let's modify our engine one more time and get more on the "radical" side. The numbers listed here are only "base" numbers. Depending on the exact camshaft position, overlap, true static compression ratio, ignition timing at idle, etc., you may find slightly different settings to work even better.

For this example, let's increase the cam to 224/234 @ .050 inch with a 112 LSA. Leaving the static compression ratio at 9:1, this modification has the engine with a relatively "weak" signal at idle. It requires more idle fuel and idle bypass air than previous examples, but your Quadrajet still does a perfect job of supplying it.

Idle tube:	.038 inch
Idle down channel:	.059 inch
Upper idle air bleed:	.070 inch
Lower idle air bleed:	.070 inch
Idle bypass air:	.106 inch
Mixture screw holes:	.099 inch
Holes in plates:	Optional*

*It is seldom necessary to use drilled throttle plates. Some Quadrajets, mostly very early production carburetors, do not use or even have the idle

Drilling small holes in the primary throttle plates adds some additional air to the engine at idle. This is usually not required if the carburetor has an idle bypass air system in place. Many early carburetors do not use idle bypass air. Drilling the throttle plates is an easy alternative to adding idle bypass air.

bypass system in place. For these models, we have the option of drilling the throttle plates in lieu of adding an idle-bypass air system. If the plates are to be drilled, do so only after testing with solid plates. Drill the holes directly in line with the idle-mixture discharge holes in the base plate through the front portion of the throttle plates.

For the last example, let's get really serious with our 350ci engine. Let's raise the static compression ratio up to 11:1 and install a relatively "radial" solid lifter camshaft with generous .050-inch specifications and a tight lobe-separation angle. We chose a camshaft with 244 degrees of duration at .050 inch tappet lift on both the intake and exhaust lobes and moved the lobe-separation angle to 108 degrees, with the intake centerline at 102 degrees. The larger lobe profiles and tighter lobe separation angle really pumps some power into the upper mid-range and high RPMs of our engine. The engine idles with a noticeable "lope" with relatively low vacuum. Note that the carburetor modifications are only slightly more generous than our

previous example. This is due to the increase in the static compression ratio as it offsets some of the lost idle quality and low-speed power from the larger camshaft.

Idle tube:	.039 inch
Idle down channel:	.064 inch
Upper idle air bleed:	.070 inch
Lower idle air bleed:	.070 inch
Idle bypass air:	.110 inch
Mixture screw holes:	.106 inch
Holes in plates:	Optional

When making idle-system modifications, we must always factor in what we believe the idle characteristics of the engine will be once placed in service. Large CID engines tolerate larger-duration camshafts with more overlap with less negative effects on idle quality. For example, a 454ci engine at 8:1 compression ratio with a 204/214 cam idles pretty much like a stock 350 engine at 8:1 compression ratio with a 195/205 duration cam. If we end up with a lot of vacuum at idle speed, then minimal idle-fuel modifications are required. A very good basic "recipe" for nearly-stock engines is as follows:

Idle tubes: .035 inch
Idle down channel: .046 inch
Upper idle air bleeds: .070 inch
(.050 inch if they are in the air horn)
Lower idle air bleeds: .070 inch
Idle bypass air: .050 inch
(none if hot air choke is used)
Mixture screw holes: .090 inch
Holes in plates: NONE

Some later-model carburetors also use larger lower air bleeds and smaller upper air bleeds. The sizes may offset each other; for example .080-inch lower air bleeds combined with .060-inch upper air bleeds ends with about the same result. If you want to follow the "recipes" listed above exactly, the large lower idle air bleeds can be resized to .070 inch in minutes.

You may also encounter pre-1970 carburetors with adjustable upper idle air-bleeds in the air horn. For these models, drill the upper air bleeds to .052 inch and use the adjustable air bleed screw as needed to fine tune the idle system.

Some early Quadrajets also use very large upper-idle air bleeds. We've seen a few Chevrolet carburetors from the early 1970s with .090 inch upper-idle air bleeds. They require quite large idle tube sizes for high-performance work, usually about .003 to .004 inch larger at every level than carburetors with smaller upper idle-air bleeds. You also have the option of driving a small aluminum plug into the bleeds and resizing them, or plugging them off and relocating the idle air bleed in the air horn. It's much easier to block off the idle air bleeds in

Once the carburetor is placed in service, the idle mixture can be fine-tuned using the adjustable idle air bleed screw as shown.

Some carburetors contain large lower idle air bleeds. They are easy to re-size using an old checkball, punch, and hammer. Place the checkball over the hole and use a flat punch and hammer to shrink the holes down. Several light taps with the hammer should be enough. Use the appropriate drill bit in a pin vise to bring them to the desired size.

Upper idle air bleeds in the main casting can be blocked using an aluminum plug made from a small piece of TIG rod or other soft material. Once we find the correct size rod, a small piece is carefully cut from it and driven into the air bleed hole. After this modification we can drill idle air bleeds in the air horn instead, or resize the plug as needed.

Drill the idle air bleeds in the air horn from the underside as shown. It is recommended to drill to a smaller size first, and then drill to final size by hand with the correct drill bit in a pin vise.

the main casting and relocate them. It can be quite challenging to plug off the upper idle air bleeds in the main casting and resize them. But, either method ends up with about the same end result.

Primary Main Fuel System

Quadrajets were delivered with many hundreds of different primary side calibrations. It would literally take an entire book to describe each and every possible jet/rod combination that was used through the entire production run of all Quadrajets. In this section we simplify things considerably by describing known combinations that work for the different models. The good news is that in a lot of cases, little if any deviation from the stock configurations is needed, especially if the engine remains stock or very close to the stock configuration, and most or all of the emission equipment is

still in place and operational. However, most have improved engine performance among their many goals for their vehicle. This usually leads to making quite a few engine modifications, increased compression ratio, larger-than-stock camshaft, aftermarket intake manifolds, headers, free-flowing exhaust, etc.

As delivered from the factory, we have seen primary jet sizes range from .064 to .078 inch and metering rods span from about .033 to near .060 inch. The type of main air-bleed system used and the size of the air bleeds themselves determine the best starting point for primary jets and metering rods in our high-performance carburetor. The main air bleeds are just as important as the primary-jet size, as they have a direct impact on how much fuel is delivered to the engine based on the jet size used with them. As we go over various sizes, keep in mind that in many cases the factory main jets and primary metering rods used are often pretty close, if they still

remain in the carburetor, and, at a minimum, provide a pretty good starting point.

The first step in determining what jetting to use is to look at what came in your carburetor and measure the main air-bleed sizes. We also need to look at what model carburetor you are working with and if it has an adjustable part throttle (APT) system in place and operational. Having APT offers an advantage as it allows fine-tuning of the part-throttle air/fuel mixtures externally once the carburetor is placed in service. We can use this to our advantage by calculating which jets and metering rods to start out with so that the portion of the metering rods resting in the jets at part throttle allows adjustments to deliver A/F ratios from rich to lean. For carburetors that do not use an APT system, we simply have to calculate the jet/rod relationships as closely as possible, as any adjustments require removing the air horn and changing parts.

In most cases, carburetors that used small main air bleeds also used small main jets. Although there are a few exceptions to this basic rule, it applies to all years and models of carburetors. In most cases, the metering rods used with any particular jet are about .030 inch smaller than the actual jet orifice size. For APT carburetors, the metering rods usually span a range so that the portion of the metering rods positioned in the jet at cruise is approximately .030 inch smaller than the jet. The basic rule for main air bleeds is that the smaller the air bleeds, the smaller the primary jet size. To simplify things, let's list a few examples so we can get an idea of how the relationships work. Then we provide some specific guidelines or "recipes" for custom calibrating our high performance carburetor.

The first example would be common for many early carburetors. Although small main jets were used the small air bleeds still provided plenty of fuel at part-, heavy- and full-throttle.

Example #1

Air bleed (main body)	.050 inch
Air bleed (air horn)	.050 inch
Main jet	.067 to .070 inch
Primary metering rod	.036 to .039 inch

The range of jets and metering rods provided simply shows that the relationship is about .030-inch difference. With carburetors that used APT, we have the ability to use a slightly larger metering rod and adjust the APT so that the tapered portion of the rod provides the most ideal part-throttle air/fuel ratio. In other words, we could install the .067-inch main jets with a .039-inch metering rod and still fine-tune the part-throttle mixtures well within an acceptable range once the carburetor is placed in service.

Example #2

Air bleed (main body)	.070 inch
Air bleed (air horn)	.070 inch
Main jet	.070 to .074 inch
Primary metering rod	.039 to .046 inch

Our second example shows larger air bleeds and larger main jets, accordingly. If your carburetor uses APT, you have the option of using the larger metering rods or metering rods with a tapered second section.

Example #3

Air bleed (main body)	.0125 inch
Air bleed (air horn)	.0125 inch
Main jet	.075 to .078 inch
Primary metering rod	.045 to .054 inch

Keep in mind that the previous examples are basically generic. The main body and air-horn main air bleeds were not always the same size. For our last example we use the later-style single main air-bleed carburetors.

Air bleed (air horn)	.080 inch
Main jet	.075 to .078 inch
Primary metering rod	.045 to .054 inch
	(.036 inch tip)

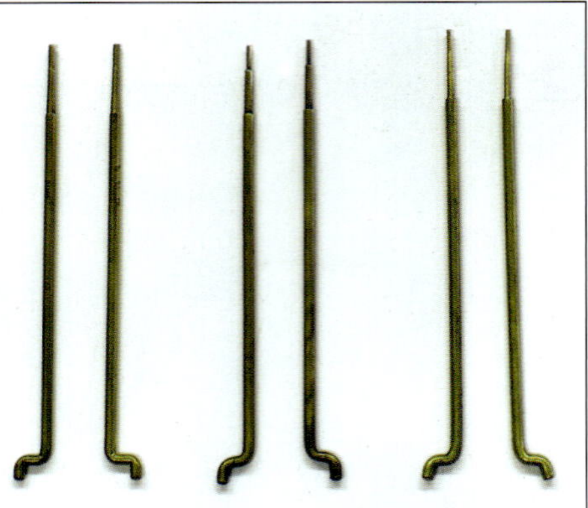

Factory primary metering rods were available in a wide selection. Early-style metering rods, shown on the right, were .040 inch longer than later metering rods. Tapered second-step rods were used for both early and late APT carburetors. Most very late units used the shorter metering rods with the .036-inch tips (far left).

The later single main air-bleed carburetors used the larger-tip metering rods with the exception of the Chevrolet units produced after 1975, which still had the early-style castings. Many of these carburetors used small main jets and small metering rods. These particular carburetors can be very difficult to tune.

Once you have calculated a starting point for your main jetting, a metering rod should be selected that ends up providing the best part-throttle calibration. For APT carburetors you do not have to use a metering rod with a tapered second section. If your metering rod's second section's largest diameter is slightly larger than the .030-inch difference, and the smallest section is slightly smaller than the .030-inch difference, you can raise the APT when tuning the carburetor to bring the tapered section between steps into the jets.

For example, if we have chosen a .073-inch main jet for a late-style APT carburetor, a .045-inch tapered second-section rod would be a good choice. For our example, we chose a rod that has about .005-inch taper from top to bottom. The highest or largest section of the rod is .045 inch, the lowest section .040 inch. The span provided by this particular metering rod allows fine metering control from rich to lean once the carburetor is placed in service. We could also have chosen a late-style single-taper rod .043 inch or larger. The single taper rod would still provide full control of the part throttle

The single main air bleed carburetors started showing up in the mid 1970s. These units were used by Chevrolet on truck engines and still used the early-style castings.

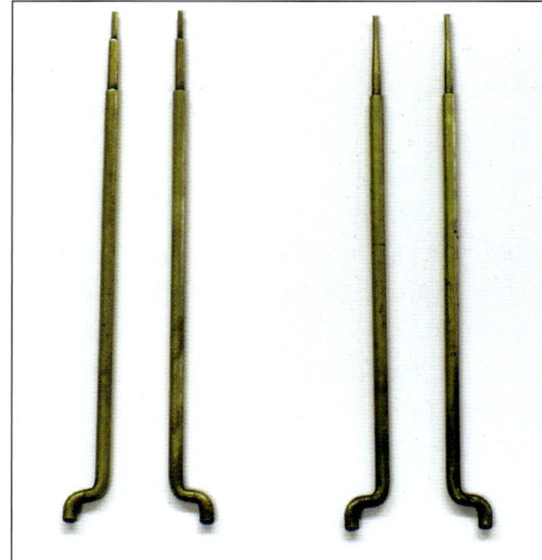

Most carburetors with APT used rods with a tapered second section (left). Two-step rods still work with APT carburetors because of the tapered section between steps (right).

A/F as the rods are raised to the tapered section between the straight .043-inch section and the .026-inch tip. The hanger must have the metering rods hanging exactly even.

Basic Rules for Primary Jetting

The first step in tuning is to select a jet size. The size of the jet controls the amount of fuel delivered to the nozzles at heavy/full throttle. The metering rod selected controls the amount of fuel delivered at light/part throttle. Changing or adjusting the metering rods only adjusts light/part-throttle fuel delivery. Changing the jet size affects both light/part- and heavy/full-throttle fuel delivery. Once you understand how the jets and metering rods affect fuel delivery, you can make needed tuning changes much more effectively.

Once the carburetor is placed into service, a few quick trips are all that is needed to verify the primary jet size. The primary jet used should provide strong, smooth power for all heavy part-throttle driving, such as climbing steep grades,

Tip in Procedure

You can use our "tip in" procedure to help verify the primary part-throttle calibration. With the engine completely warmed up, set the speed via the fast-idle or idle-speed screw to 2,000 to 2,200 rpm. Gently "tip in" the choke flap or partially cover the choke housing if the choke is not used. Don't close the flap enough to choke out the engine, but just enough to richen up the mixtures a bit. You should see an engine speed increase of about 50 to 100 rpm. This tells you that your primary settings are just on the edge of lean. If there is no engine-speed increase or RPM fall off some, pull a small-manifold vacuum hose from the carburetor. If

The "tip in" procedure is used to set the APT on this late-model carburetor.

the engine speed increases due to the additional amount of incoming air, the mixtures are a bit rich. If you are using an APT carburetor, set the metering rods to provide a very slight RPM increase during the "tip-in" procedure. Verify the settings by pulling off a small manifold vacuum hose; the engine RPM should stay about the same or the engine should slow very slightly. This procedure provides a good starting point. You can now place the carburetor in service and do some real-world testing. Future metering changes may be needed, as some vehicles prefer slightly rich or slightly lean settings for best overall efficiency.

passing, etc., without employing the secondaries. If the engine surges or hesitates, it is most likely a bit lean. If acceleration is "flat" and you see any black smoke out the back, the mixture is rich. In either case if acceleration is not smooth and strong at heavy part-throttle, change the jet size to achieve the desired results. Once you have established the correct jet size, change the metering rods accordingly.

Power Piston Springs

The power piston spring is also a tuning tool for the primary side of your carburetor. Even so, the power piston spring chosen is not going to work well unless the primary-jet and metering-rod relationship is correct. The power piston spring's only purpose is to raise the metering rods during low-vacuum situations. The spring length, rate, and number of coils determine exactly when this occurs, as vacuum falls off during heavy

power piston springs to raise the metering rods. The stronger springs enhance throttle response during these situations.

Accelerator pump

As described in Chapter 2, the accelerator pump is used to supply large amounts of fuel on demand with quick throttle movements. The accelerator pump system on the Quadrajet is superb and needs very little help. The exit holes

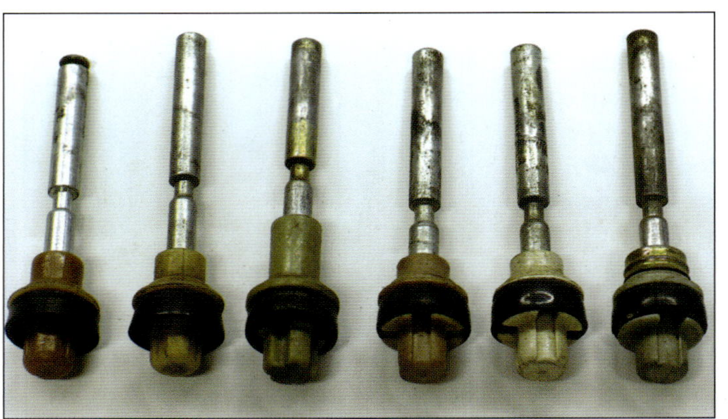

Quite a few different accelerator pumps were used. Here is a sample of used pumps removed from different year and model carburetors.

in the RPM range. A good starting point for high-performance carburetors is .028 inch. The largest size that we've ever used was .036 inch. Increasing the accelerator pump discharge hole size should be done in small increments after considerable testing. The primary metering system should be fine-tuned first, as the accelerator pump is designed to complement the primary side for quick-acceleration situations, not for normal light-throttle operation.

Factory power-piston springs were set up exactly for the application. There were quite a variety of springs used.

In later years a spring was used behind the seal to keep it in contact with the accelerator pump bore. These pumps are preferred for high-performance use.

and full-throttle driving. The basic rule for power-piston spring selection is to use the strongest spring that does not raise the power piston at idle speed. Off-idle during light-throttle cruising always has enough vacuum to keep the power piston fully seated. Any quick, heavy, or full-throttle operation lowers the vacuum in the engine to cause any of the

in the air horn can be fine-tuned for particular applications. From the factory, the holes are typically very small, around .026 inch or smaller. Most factory vehicles had a poor power-to-weight ratio, and the smaller holes provide a longer duration of fuel delivery. For high power-to-weight ratio cars, you may want to increase the size for more fuel earlier

You can also fine-tune pump delivery by shortening the accelerator pump shaft. There were quite a few different pumps used through the years of production.

For high-performance use, the late-model short-pump with garter spring is the best choice. These can be obtained from late-model carburetors or purchased separately from high-performance warehouses or auto parts stores. They also used a spring to help keep the pump-cup seal in contact with the bore for improved pumping efficiency.

The blue or red pump seals work best. They are more resistant to swelling from modern fuels. Most of the black rubber or "Buna" seals do not last long after exposure to today's modern gasolines.

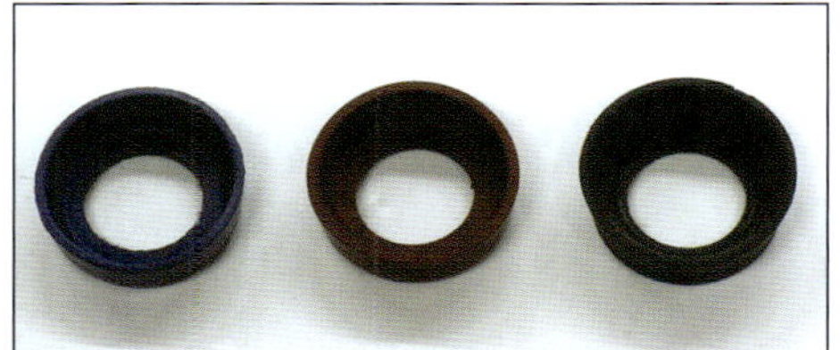

Pump seals were made from several different materials. The blue seals are the most resistant to swelling, shrinking, drying up, and cracking. Some modern fuels are very hard on the black rubber seals. We've seen them swell up and fall off the pumps in less than 30 minutes after being exposed to fuel!

Fuel Inlet Seats

The fuel-inlet seat is a very important tuning tool for high-performance Quadrajet carburetors. Quadrajets have long been rumored to have insufficient fuel capacity for serious high-performance use. This rumor comes from the poor float-fulcrum design used by the early units. The early units had insufficient leverage on the fuel-inlet needle/seat assembly and are very sensitive to large inlet-seat diameters and/or small floats. This fact does not mean that the early units can't be used for high-performance work. They simply need a lot of fuel available to them at relatively low pressure.

Later Quadrajets having the float fulcrum moved forward can utilize a smaller float and a larger fuel-inlet seat orifice size. Few rebuild kits show up these days with the larger seats. They are

The factory fuel inlet seats were available in a wide variety of diameters. Most rebuild kits contain fuel-inlet seats with a .125-inch hole diameter. This is marginal for high-performance use. The fuel inlet seats in the picture are right to left: .110 inch, .125 inch, .130 inch, and .135 inch.

typically between .100 and .125 inch. The .125-inch seat is fine for mild to medium high-performance engines to about 350 hp. The .125-inch seat is a good choice for general street use, providing good fuel flow into the carburetor without any flooding issues related to high-pressure, hard-stopping, quick vehicle maneuvering. The larger .135-inch seat is about as large as we can utilize without a fuel-pressure regulator. They hold fine with the later float designs to at least 6.5 psi. Again, the key to keeping the carburetor full is volume, not fuel pressure. The large .149-inch fuel-inlet seats are available through the aftermarket. The .149-inch inlet seat should be used with a regulator. With the small float they hold to at least 6 psi and are a good choice for large-CID, high-horsepower applications.

The rule of thumb for selecting a fuel-inlet seat is to use the smallest diameter that works with your particular

Solid fuel inlet seats started showing up in the mid 1970s. The factory discontinued "windowed" fuel inlet seats, claiming improved fuel control with the solid version. Some aftermarket rebuild kits contain drilled fuel-inlet seats. Our testing has shown that any type works equally well in a high-performance carburetor. The only advantage provided by the solid seat is that it does not allow the carburetor to drain back as low in the fuel bowl. This can be an advantage if drainback is causing a hard starting problem after the vehicle sits for any length of time.

application. We have found through direct testing that the fuel system is just as important as fuel inlet size and fuel pressure. Most factory vehicles have a fuel system designed for far less power than most custom high-performance engines produce. The pick-up location in the tank is usually closer to the front than the rear and there are always quite a few bends in the stock lines as they route through the frame. Mechanical pumps are quite sufficient at pumping fuel in a static environment. We've had no trouble supplying powerful engines on the dyno with stock mechanical pumps, but found that in most cases we could not get them to keep the carburetor full on the street or track under hard acceleration. One has to consider that when you are at full throttle in low gear and pinned against the seat, most of the fuel in the tank is pinned against the gas cap! At this point, it becomes increasingly difficult for a mechanical pump to pull the fuel through nearly 14 feet of fuel line and deliver it to the carburetor in sufficient volume to keep it full.

For most high-performance applications, the .135-inch fuel-inlet seat is the best choice. With a good fuel system it is seldom necessary to go any larger for most applications. Some fuel-inlet seats are "windowed" or have multiple round

The .149-inch fuel inlet seat is available through the aftermarket. These seats are great at letting large amounts of fuel into the fuel bowl, but they can be problematic with some models and high-pressure aftermarket fuel pumps.

Custom Fuel-Inlet Seats

Making our own high-performance fuel-inlet seat takes just a few minutes. Drill the seat with the desired size drill bit from the bottom side.

The drilling procedure may leave slight imperfections on the edge of the hole. Gently form a new seat with a quick tap using an old checkball, punch, and hammer as shown. The seat should be resting on a soft surface to prevent damage to the threads.

It is quite easy to make a custom fuel-inlet seat. This requires an extra accelerator pump checkball, punch, drill press, correct drill bit, and a small hammer. Invert a smaller fuel-inlet seat and carefully drill it through the center with the desired bit size. Make sure the drill bit table is exactly 90 degrees to the chuck.

Set the fuel-inlet seat upright on a solid surface, drop checkball onto the seat, and give it a couple of quick taps with the punch using a small hammer.

This forms a new seat. Use a vacuum tester to test the new needle/seat assembly before placing it in service. The assembly should show zero leakage for at least five minutes. If the assembly fails the vacuum test, carefully clean the seat and the tip of the needle and re-test. It may be necessary to debur the seat assembly by running the drill bit through and re-seating the ball onto the seat.

Always vacuum test the needle-and-seat assembly prior to placing them in service. It is highly recommended that all fuel inlets seats be "staked" with a checkball prior to use. This forms a wide flat seating surface and improves sealing and longevity of the needle valve.

holes drilled in them just above the seating surface.

This provides additional paths for fuel to enter the fuel bowl. The factory discontinued the use of the "windowed" fuel inlet seats in the mid 1970s. The goal was to improve fuel control as it enters the fuel bowl. Lacking the window or drilled holes also reduces the possibility of fuel bowl drainback after engine shut down, an occasional problem with Quadrajet carburetors. We've done extensive dyno, street, and track testing with various types of solid and windowed fuel-inlet seats. Since the fuel pressure is at zero past the hole under the needle valve, having the windowed or drilled seats shows no performance advantage that we have been able to detect.

Floats

Floats are available for all models in brass and nitrophyl.

The brass floats lack the buoyancy of the OEM material. The downside of the nitrophyl floats is that on occasion they soak up some fuel and sink, causing increased fuel-delivery and flooding issues. The nitrophyl floats are recommended for high-performance applications and should be closely inspected prior to use. They should be rejected if any surface defects are discovered. Always replace any used float with a new one regardless how good it looks.

Small-diameter floats are available for the early center-fulcrum models. They are usually okay with stock low-pressure fuel pumps up to seat diameters of .135 inch. A fuel-pressure regulator is highly recommended if they are used with aftermarket high-pressure mechanical or electric pumps.

The 1969–1974 non-Oldsmobile carburetors can be outfitted with the later-style floats used in the 1975 and later carburetors. The smaller float takes up less room for fuel and still provides sufficient leverage on the inlet needle to

Small floats are available as replacements for the early-style pre-1969 carburetors. The smaller floats allow more fuel in the fuel bowl. They can be troublesome with high-pressure fuel pumps and large fuel inlet seats. A fuel-pressure regulator may be needed to use the smaller float, but the additional fuel available in the fuel bowl helps these carburetors when used in high-performance applications.

control the fuel level. As with the early-style float arrangements, a fuel-pressure regulator is highly recommended.

There were several different floats used in the 1975 and later carburetors. The smaller floats combined with a .135-inch fuel inlet seat are a good choice for most street/strip applications. The larger

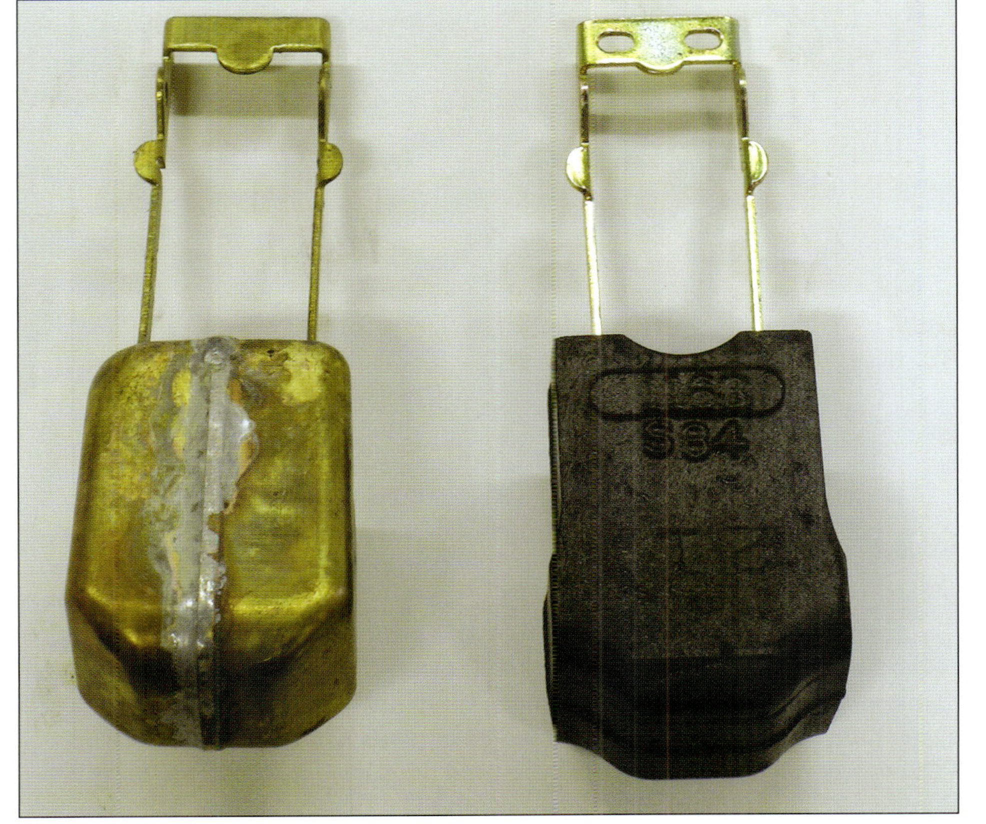

The factory used nitrophyl floats. They provide excellent buoyancy and fuel control. Modern fuels can be especially hard on them and cause them to get heavy and sink. They should always be replaced during rebuilding. Aftermarket brass floats are available for most models. The brass floats are impervious to fuels. Occasionally they have a leak at one of the soldered joints. They should always be submerged in warm water prior to using them. Any leak shows up as a stream of bubbles.

The float on the left was used from 1969 to 1974 for most models. The smaller float on the right was used in the later-style carburetors produced in and after 1975. The smaller float works in the early-model carburetors and provides some additional fuel-bowl capacity. A fuel-pressure regulator may be needed.

inlet seats can be used, but a fuel pressure regulator is recommended.

Pullover Enrichment System

The primary side pull-over enrichment system was used throughout the years of production and can be employed on most models. It provides no real performance improvement. Fuel is only added with heavy part-throttle operation to compensate for lean primary side calibrations. It does not hurt anything if already in place. Even so, direct street and track testing has shown no overall performance improvement by adding the system.

The secondary enrichment system is used to help get the huge secondaries on line without stumble, bog, or hesitation. Make sure the two small holes entering the fuel wells are at least .035 inch. On some models they are as large as .040 inch. Their only purpose is to restrict fuel to the wells so the fuel does not continue to flow once the main system is operational.

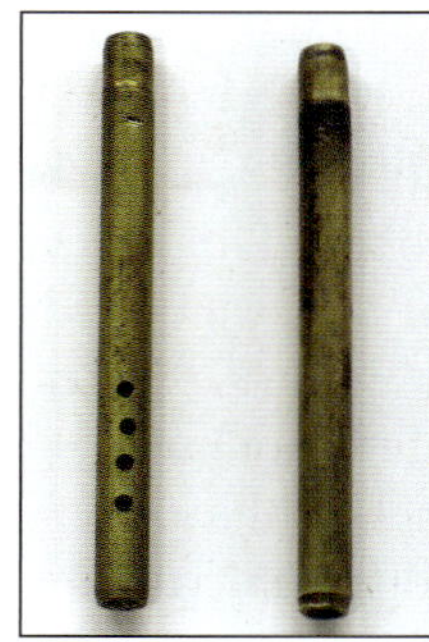

A common "high-performance" modification was to drill a series of small holes in the brass tubes that supply fuel to the secondary accelerating system. Do not drill the tubes, as this will cause the fuel to be quickly sucked out of the wells, exposing the small holes to air. Solid tubes are required for full fuel delivery from this system.

Going beyond .040 inch is seldom necessary and can affect the overall air/fuel ratio at full throttle at higher RPM.

Do not under any circumstances drill holes in the brass tubes that hang into the fuel wells. As soon as the fuel level drops below the first hole, it begins to draw air instead of fuel and affects fuel delivery to the engine.

Here are two different floats used in the 1975 and later carburetors. The small float is preferred because it allows more fuel into the fuel bowl.

Late-model carburetors that used the primary pull-over enrichment (POE) system are easily identified by the six brass tubes hanging from the air horn. POE was rare in the later units as they had an excellent adjustable part-throttle (APT) system for fine metering control.

Some carburetors came from the factory with the secondary accelerating system holes exiting under the air flaps. This is the most desirable exit location since fuel has a direct path to the engine at the initial application of the huge secondary throttle plates.

The exit holes in the air horn should be exposed to the underside of the secondary air flaps. This provides a direct shot of fuel at the initial opening of the secondary throttle plates. Some Quadrajets had the holes located under solid flaps.

Many early Pontiac carburetors had huge slots in the air flaps and large openings in the air horn. Although rumored as a superior setup, the slots in the flaps are so large that they allow too much air to enter the engine before fuel begins to flow.

This arrangement was discontinued in the mid 1970s. Pontiac relocated the holes directly in line with solid flaps so that a portion of the holes was exposed to the underside. This proved to be a much better arrangement.

Most Quadrajets have small exit holes located over solid flaps. The easiest modification for these carburetors is to simply "notch" the flap to expose the holes to the underside. Small notches are all that is

In lieu of using huge slots in the flaps, Pontiac went to solid flaps with large openings directly in line with them. This exposed a portion of the holes to the underside of the flaps and improved fuel delivery to the engine when the secondaries are first applied.

Early Pontiac carburetors used huge slots in the secondary air flaps. This allowed fuel a direct path from the accelerating system to the engine. The flaps were discontinued in the early 1970s after factory discovered that the flaps allowed too much air to enter the engine.

Most Quadrajets have small holes located over the secondary air flaps. This is the worst scenario and can cause the engine to hesitate, stumble, or bog when the secondaries are first opened.

A small "notch" can be added to the air flaps to expose the fuel holes to the underside of the flaps. This modification is easier than relocating the holes under the flaps.

Exit Hole Relocation

If class racing prohibits modifying the air flaps, the holes can be relocated under the flaps. Remove the brass fuel supply tubes. Melt a few small pieces of solder into the holes and drive them in far enough with a small punch to seal off the original holes. If too much solder was used to seal off the old holes, simply drill in with a drill bit until the desired distance is reached and the old holes are still sealed off. Use a small piece of wire to measure the depth of the holes so drilling under the flaps intersects the holes. Use a sharp punch to locate new holes just under the air flaps and carefully drill them into the fuel-supply passage. Use a drill bit slightly smaller than the desired finished diameter. Finish drilling slowly by hand using a pin vise. Test the new supply holes with compressed air.

The secondary accelerating exit holes can be easily relocated under the flaps. Use a sharp center punch to locate the holes in line with the fuel passage. A small drill bit can be put into a pin vise to provide drilling access to the holes. It is recommended that the holes be drilled to a smaller size first, and then final-sized by hand with the correct drill bit in a pin vise.

needed as the fuel runs nearly straight down the leading edge of the secondary bores when the throttle is quickly opened.

Secondary Emulsion Tubes (Air Bleeds)

The two inner brass tubes add some air to the main fuel supply to the secondary fuel nozzles. Some air horns have two additional holes in the casting beside the brass tubes. Early air horns that used pull-over enrichment use larger tubes with no restrictions. For the early-style POE carburetors, do not modify the tubes or exit holes into the primary bores.

Some very early models used pull-over enrichment with larger brass tubes in the main fuel-supply passages.

For all other models, drill the holes in the brass tubes to .036 inch. Block off any additional holes in the air horn beside the brass tubes. This can be accomplished by driving in small homemade plugs from aluminum-tip rods, staking carefully both sides to close off the holes or using Marine Tex epoxy.

Check to make sure the factory fuel-supply nozzles are not obstructing the fuel-supply holes in the air horn. It's common for the nozzles to be driven in

Some air horns use an additional pair of drilled holes next to the small brass tubes that extend into the secondary fuel passages. These holes should be blocked off for high-performance applications. The material around them can be staked with a small punch to close the holes. A small dab of epoxy over the holes works equally well.

It's common for the secondary fuel nozzles to be driven in too far and close to part of the fuel passage in the air horn. A small stone on a rotary grinder can be used to remove the exposed portion of the tubes, or they can be pulled and driven back into the correct location.

against the brass tubes or far enough in to obstruct fuel delivery.

A small die grinder and stone can be used to open up the passages. Another alternative is to carefully remove the tubes and grind the leading edge at an angle so it doesn't inhibit fuel flow when driven back in. Make sure to deburr the tubes after grinding. Use red Loctite and stake the fuel-supply nozzles in place any time they are removed for modifications.

Secondary Air Flaps, Shaft, Hanger, and Metering Rods

The secondary air flap shaft should move freely its entire distance from fully closed to fully open. The full-open angle varies considerably for different carburetor models and applications. Grind most of the stop from the end of the linkage and drill/tap the air horn to install an adjustment screw.

Most of the stop on the linkage can be ground off and an adjustable stop made by drilling and tapping the air horn. Shown is a #3-48 screw used to adjust the open angle of the secondary air flaps.

This allows full range of adjustment for our high-performance carburetor. Always inspect the plastic secondary cam that raises the metering rods. It may be worn or cracked. Worn cams are one of the most common reasons for poor results when trying to tune a Quadrajet carburetor, followed closely by a worn or broken return spring.

Always inspect the secondary air-flap spring and the plastic cam. Neglecting to replace these worn items is a big reason why many Q-jet carburetors do not respond to high-performance modifications.

The selection of factory metering rods is somewhat limited. Currently, only .030-, .041-, .052-, .056-, and .066-inch metering rods are available in the aftermarket. Access to a small lathe may be needed if a precise assortment of metering rods is needed between available sizes.

Secondary hangers have a letter code indicating where the metering rods are positioned on the secondary disks in the bottom of the fuel bowl. Metering rods vary in taper, tip diameter, and tip length. They were given a two-letter designation by the factory (see Appendix A).

Metering rods stamped AX or CE are a good starting point. These rods feature small .040 to .041-inch-long tips and a rich taper. CE rods are still available from the aftermarket through high-performance parts suppliers. Most late-model quadrajets were delivered with lean secondary metering rods. We have to obtain richer rods or make them from the larger cores. A small bench lathe can quickly turn down the larger cores into any tip size and length. Good results can also be obtained by using a ⅛-inch drill clamped in a vise as a lathe and carefully grinding down the tips with a jeweler's file. The tips should be blended into the taper so the rods do not hang up on the disks in the fuel bowl.

The secondary air-flap adjustment spring should be set ½ to ¾ turn clockwise after it just closes the air flaps. Check the shaft for full movement from

Measure the distance from the leading edge of the secondary air flap when fully open, down to the leading edge of the rear of the opening in the air horn as shown. This provides a baseline for setting the secondary flap full-open position.

metal link. In addition to opening the choke flap on cold starts, it also prevents the secondary air flaps from opening too quickly when we go to full throttle. The choke pull-off can be used as a tuning tool in lieu of omitting it and having to wind the secondary air-flap spring too tight to compensate. By calibrating pull-off release time, we can use less spring tension and make our high-performance Quadrajet progressive in secondary operation. Calibrating the choke pull-off prevents any lean spots and helps get the huge secondaries on line seamlessly. By improving the pull-over system and using the correct rods and hanger height, the end result is flawless full-throttle transition from idle to the shift point.

High horsepower-to-weight ratio vehicles can utilize a very fast choke pull-off. Lower power-to-weight applications require slower opening rates. Most late-model choke pull-offs used a very slow opening rate, typically three to five seconds. This proves to be too long for most high-per-

The link from the choke pull-off to the secondaries helps control the opening rate of the air flaps. It is not recommended to remove the link. This is because it often requires too much spring tension to keep the flaps from opening too quickly. Modify the choke pull-off instead for the correct release time for the application.

closed to open. With the flaps fully open, measure the distance from the leading edge of the flaps back to the edge of the rear of the opening in the air horn. The maximum open distance should be 1.270 inches. This duplicates the angle used by Edelbrock for its "850" cfm carburetors. For most applications, 1.280 to 1.330 inches is a good starting point.

Choke Pull-Offs

The choke pull-off is connected to the secondary air-flap linkage via a ⅛-inch

Any choke pull-off with a bent vacuum supply tube can be modified for access to the restriction. Most have to be shortened some and heated up to straighten the tube.

Making Choke Pull-Off Restrictions

Another alternative to modifying the choke pull-off is to remove the restriction in the choke pull-off supply tube entirely and use custom restrictors placed inside the vacuum supply hose to the pull-off. Restrictors can be made from soft aluminum TIG rods or even a small section from a plastic golf tee, or most any soft plastic material of similar diameter. Plastic proves to be easier to drill with the tiny drill bits. If metal restrictors are used, it is easier to drill them clear through with slightly larger bits first, then peen the hole nearly shut and size with the smaller bits. Armed with a handful of custom-made supply hoses, you can cut to the chase quickly with your full-throttle tuning.

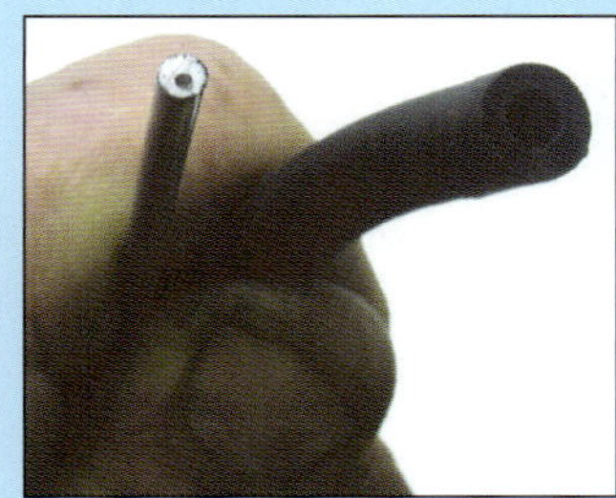

An alternative to drilling out the restriction in a choke pull-off is to remove the restriction. Custom-made restrictions can be fabricated and inserted into the vacuum-supply hose. Several different sizes can be made up ahead of time for custom tuning on the street or track.

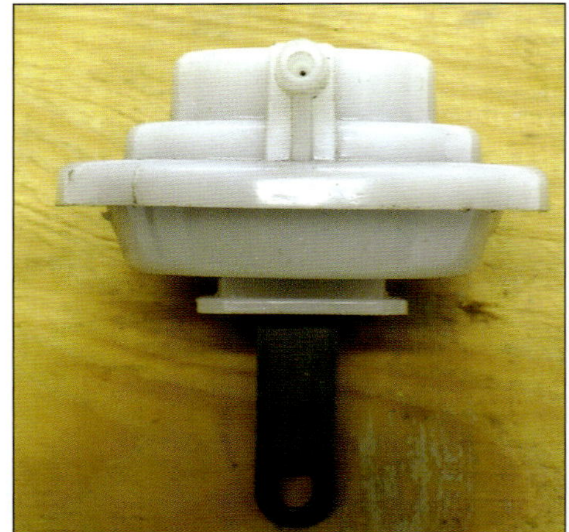

Most early carburetors used a small plastic choke pull-off with a straight tube. These are easy to modify for release time.

Straight tube choke pull-offs are available for most late-model carburetors. They provide much easier access to the restriction in the housing to adjust the release time of the pull-off.

Using a tapered reamer in lieu of tiny drill bits to open up the restrictions in choke pull-offs offers more flexibility than drilling with sized drill bits.

formance applications. You can adjust the opening rate by accessing the restriction in the vacuum supply tube. Some choke pull-offs offer easy access to the restriction in the vacuum supply tube, while others do not.

Most early carburetors can be outfitted with the plastic pull-off having a straight tube. These are inexpensive and very easy to modify.

Some models used a steel housing pull-off. If the supply tube doesn't offer access to the restriction it can be gently heated and straightened.

You may need to shorten the tube for the tiny, short drill bits. This is because the restriction is usually at the base of the supply tube. The hole size affects the opening rate. Most canisters require hole sizes between about .016 inch and .022 inch. Test the opening time by fully compressing the pull-off linkage or using a vacuum tester. Release the lever or vacuum and count in seconds how long it takes the linkage to return to the fully extended position. Around two seconds is a good starting point for most applications.

Recipe for Success

Now that you have covered modifications to specific areas of your high-performance carburetor, let's look at some specific combinations of parts and settings. These provide baselines for performance and save many hours of testing and tuning.

The first "recipe" is a good starting point for mild applications to about 350 hp. The cams used with mild engines should supply at least 14 inches of vacuum at idle speed. These engines should make peak horsepower below 6,000 rpm and have strong torque in the mid-range. Make sure to note that there should be only one pair of upper idle air bleeds, either in the main body or the air horn. This calibration works for models using two pairs of main air bleeds or single main air bleed units as noted.

The second recipe is for engines where large enough cams have been installed to lower the vacuum at idle to about 10 to 14 inches. These engines typically "lope" a bit at idle and need additional idle fuel for good idle qualities in

Continued on page 115

First Recipe

Idle tube:	.036 inch
Idle channel:	.046 inch
Lower idle air bleed:	.070 inch
Upper idle air bleed (main body)	.067 to .070 inch
OR	
Upper idle air bleed (air horn)	.050 to .052 inch
Accelerator pump discharge holes:	.026 to .028 inch
Main air bleed (main body)	.070 inch
Main air bleed (air horn)	.070 inch
OR	
Single main air bleed carb:	.045 to .060 inch
Main jet*:	.071 to .073 inch
Primary metering rod:** (for APT carbs use a tapered rod at least .029 inch less than the main jet at the widest point)	Approx. .030 inch less than jet size used
Fuel inlet seat:	.125 to .130 inch
Float level:	¼ inch
Secondary POE well restriction:	.032 to .038 inch
Secondary POE restriction:	.052 to .055 inch
Secondary tube restriction:	.036 inch
Secondary hanger:	K or higher
Secondary metering rods:	.045-.055 inch
Air flap open distance:	1.300 to 1.330 inches
Secondary flap spring:	¾ to ⅞ turns
Choke pull-off release time:	2 to 3 seconds

* For late-style single main air-bleed carburetors with .060 to .080 inch main bleeds, use .074- to .076-inch main jets
** For single main air-bleed carbs use a 50, 52, or 54 M primary metering rod

Large main air bleed carburetors

Notice that all listings using two pairs of main air bleeds use the same sizes. Chevrolet released quite a few carburetors with huge main air bleeds, near .125 inch. If your carburetor has the large main air bleeds, all of the modifications listed for each example still apply, except for the main-jet and metering-rod choices. Increase the main jets to .076 or .077 inch. Use a metering rod about .030 inch smaller as a starting point. The metering rods must have the smaller .026-inch tips. If a tapered second-section rod is used, make sure the tapered section spans a few thousandths of an inch in either direction of the .030-inch difference.

Another option would be to borrow four brass restrictions for a donor carburetor sized to .070 inch. You can then follow the recipes exactly. You can also make main air-bleed restrictions from a small piece of .125-inch aluminum TIG rod.

A .077-inch main jet is shown beside a 49K metering rod. The 49K metering rod has a tapered second section ranging from .049 inch down to .043 inch. It also has a .026-inch tip, required for use in the large main air bleed carburetors.

A small section of a TIG rod can be fabricated into a main air bleed. Using a fine hacksaw blade, cut a small piece from the TIG rod. It can then be coated with Loctite and carefully driven into the air-bleed hole. Once the Loctite has set, carefully drill to the final size desired.

Second Recipe

Idle tube:	.037 to .038 inch
Idle down channel:	.052 to .055 inch
Lower idle air bleed:	.070 inch
Upper idle air bleed (main casting)	.067 to .070 inch
OR	
Upper idle air bleeds in air horn:	.050 to .055 inch
Accelerator pump discharge holes:	.028 to .030 inch
Main air bleed (main body):	.070 inch
Main air bleed (air horn):	.070 inch
OR	
Single main air-bleed carbs:	.080 to .086 inch
Main jet:*	.072 to .074 inch
Primary Metering rod: **	Approx. .030 inch less than jet size used
(for APT carbs use a tapered rod at least .029 inch less than the main jet at the widest point)	
Fuel inlet seat:	.135 inch
Float level:	¼ inch
Secondary POE well restriction:	.035 to .040 inch
Secondary POE restriction:	.055 to .059 inch
Secondary tube restriction:	.036 inch
Secondary hanger:	G or higher
Secondary metering rods:	.036 to .048 inch
Air flap open distance:	1.280 to 1.300 inches
Secondary flap spring:	½ to ¾ turns
Choke pull-off release time:	1.5 to 2 seconds

* For late-style single main air-bleed carburetors use .074- to .077-inch main jets
** For single main air-bleed carbs use a 50, 52 or 54 M primary metering rod

Third Recipe

Idle tube:	.038 to .040 inch
Idle down channel:	.059 to .064 inch
Lower idle air bleed:	.070 inch
Upper idle air bleed (main casting)	.067 to .070 inch
OR	
Upper idle air bleeds in air horn:	.052 to .059 inch
Accelerator pump discharge holes:	.028 to .033 inch
Main air bleed (main body):	.070 inch
Main air bleed (air horn):	.070 inch
OR	
Single main air-bleed carbs:	.085 to .090 inch

Continued on Page 114

Third Recipe Continued

Main jet:*	.073 to .075 inch
Primary metering rod:** *(for APT carbs use a tapered rod at least .029 inch less than the main jet at the widest point)*	Approx. .030 inch less than jet size used
Fuel inlet seat:	.135 to .149 inch
Float level:	¼ inch
Secondary POE well restriction	.038 to .040 inch
Secondary POE restriction:	.059 to .064 inch
Secondary brass tube restriction:	.036 inch
Secondary hanger:	G or higher
Secondary metering rods:	.033 to .045 inch
Air flap open distance:	1.240 to 1.300 inches
Secondary flap spring:	¼ to ½ turns
Choke pull-off release time:	1 to 2 seconds

** For late-style single main air-bleed carburetors use .076- to .078-inch main jets*
*** For single main air-bleed carbs use a 52 or 54 M primary metering rod*

A small amount of solder or lead can be melted into the power piston bore to seal off the vacuum supply.

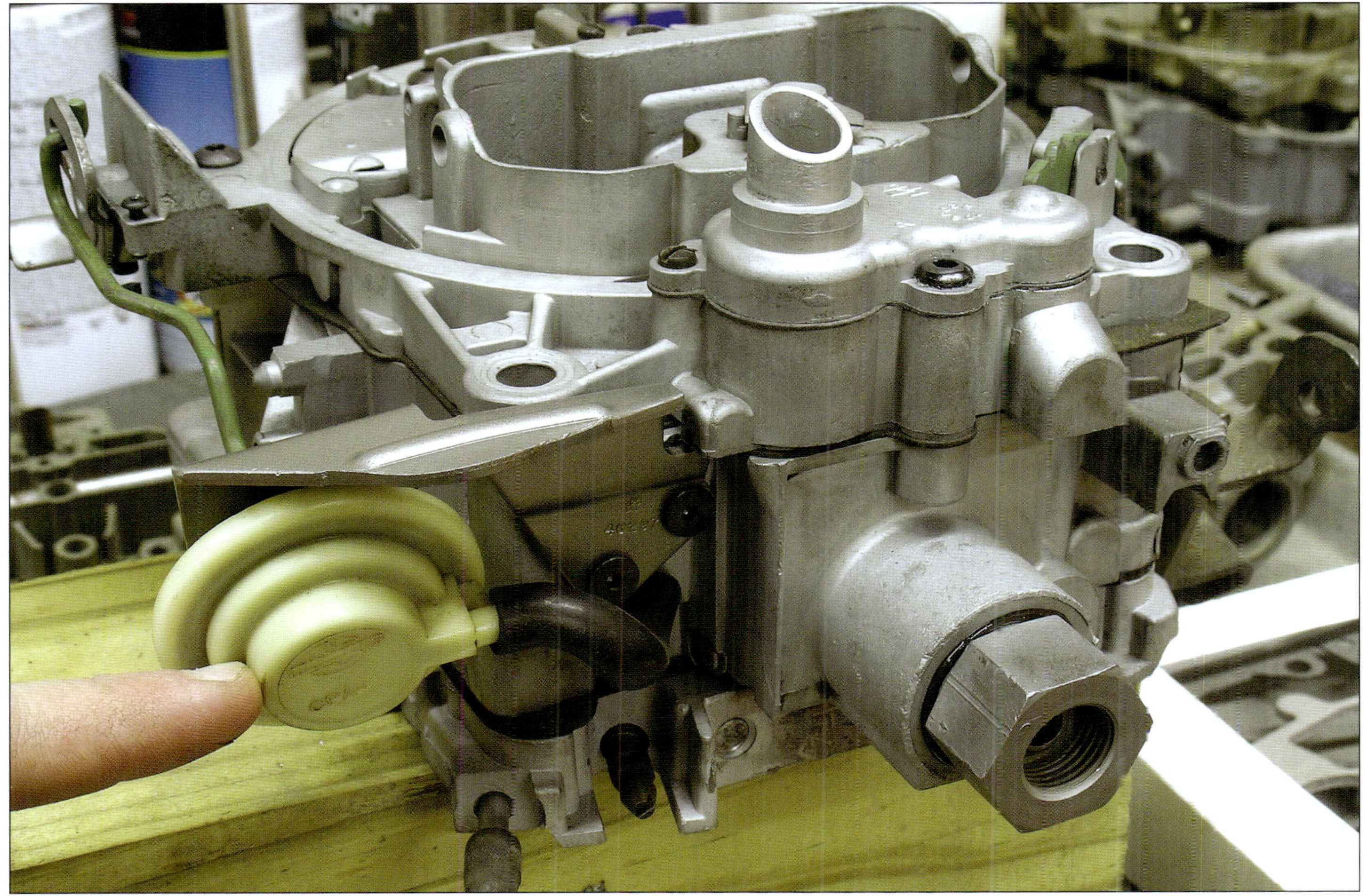

For high-performance applications, some later-model carburetors can be converted to use the early-style plastic choke pull-offs. The plastic choke pull-offs are inexpensive and easy to modify.

and out of gear. The higher horsepower output requires quicker settings and typically a bit more fuel right off idle.

The third recipe is for serious high-performance street/strip engines where the vacuum at idle ranges from about 8 to 12 inches. These engines typically have higher compression ratios than "stock" engines, more aggressive cam profiles, looser torque converters, and taller gearing.

All of the recipes listed should be used as starting points. In cases where the engine would fall in between the descriptions given, stay on the conservative side. You can always go back and make changes after initial testing.

The calibrations are also intended for sea level to about 2,000 feet. At higher altitudes the main jet and rod choices should be reduced accordingly. Reduce the main jet size .001 inch from above 2,000 feet to about 3,000 feet, and .002 inch at 4,000 feet and higher elevations.

Full-Race Carburetors

The Quadrajet makes an excellent racing carburetor. Once you have upgraded your carburetor and fuel system to keep the carburetor full under all conditions, you can select a core and modify it as needed for excellent results at the track. For most applications, you can retain the power piston and primary metering rods. For drag racing, it can be beneficial to remove the power piston and primary metering rods. This improves starting-line performance if the driver is going to come up hard on the

torque converter or if a trans-brake is used. Any transition between cruise and heavy full-throttle metering is eliminated. When the driver comes up on the converter or trans-brake, there is no tendency for the car to shift in the light due to a change in the A/F ratios. If the power piston and primary metering rods are omitted, reduce main jet size by approximately .006 to .008 inch for carburetors using .070 inch /.070 inch main air bleeds.

Remarkably, leaving out the power piston and metering rods does not have an adverse affect on part-throttle performance. The only negative is that the engine is just a tad rich when cruising around in the pits.

The choke pull-off and link to the secondary air flap shaft should still be

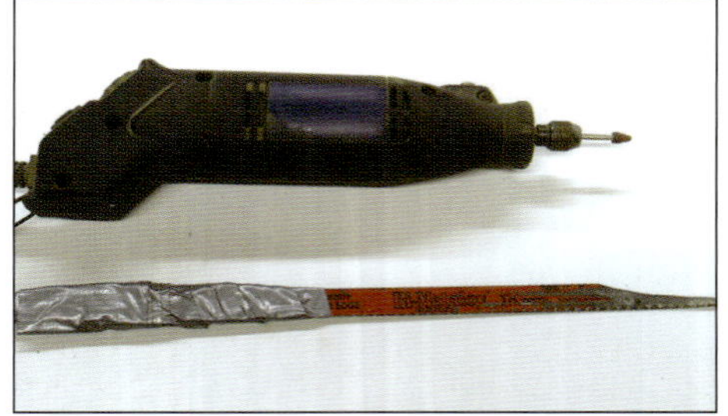

The hacksaw blade shown has been modified for easy access to removing the outer booster rings. A tiny stone in a rotary tool is used to remove any material left in the carburetor after removing the booster rings.

retained, even for race-only applications. For cars having superior power-to-weight ratios, the choke pull-off release time should be set accordingly. With these vehicles you can run very little spring tension on the air flap shaft, and use a quick release pull-off to control the opening rate. Omitting the choke pull-off often mandates winding the air flap spring too tightly, and this can affect full-throttle fuel air and fuel delivery.

Removing the Primary Booster Rings

Removing the primary booster rings increases the CFM capacity of the carburetor. They are relatively easy to remove, but the carburetor has to be completely recalibrated. Removing the booster rings eliminates one of the venturi areas. This reduces the signal, or pull, to the main jets. We have to increase the main jet size by .001 to .002 inch and use a primary metering rod closer to .035 to .038 inch less than the jet size. For most street and street/strip applications there is no real advantage from this modification. If our engine requires more CFM than the carburetor we are using is capable of, then there may be a performance improvement from this modification. For marine, towing, and large-CID engines, there can be advantages to having more

Using our custom hacksaw blade with the carburetor held by soft jaws in a big vise, it only takes a few minutes to remove the booster rings.

CFM available on the primary side of the carburetor. Since marine and towing applications spend a lot of the time with the engine heavily loaded at an increased throttle angle, this modification can be very beneficial.

Removing the outer booster rings is one modification that permanently alters the physical characteristics of the carburetor. It is recommended that a used low-cost core be obtained for this modification. The tools required are a modified fine-tooth hacksaw blade and a small rotary grinder.

Once the rings are removed, remove any additional material left in the bores that might block airflow. After the casting is cleaned up, use a small sanding roll or polishing stone to smooth and blend the contour of the bores.

The final product, after removing the booster rings. A tiny stone and sanding roll were used to smooth the casting, followed by bead blasting with fine abrasive glass beads.

EDELBROCK Q-JETS

For decades the Edelbrock company has made aluminum heads, intake manifolds, carburetors, and other accessories for the high-performance market. It offered both replacement and high-performance Quadrajet carburetors for many years. The replacement units were limited to several basic styles, all with side fuel-inlet locations and both electric and divorced chokes.

For high-performance use, Edelbrock offered their "RPM" Q-jet. It was the first and only true high-performance Quadrajet ever offered in large production by an aftermarket manufacturing facility. Based on the late Chevrolet castings, they featured a host of upgrades to the standard units. Edelbrock rated them at 850 cfm for powering engines over 450 hp.

Edelbrock installed a high-performance fuel inlet seat with a hole diameter of .149 inch. The accelerator pump had a shortened shaft and stiff duration spring for maximum fuel delivery. They featured a stiff spring under the pump to help return the throttle to the closed

The only aftermarket high-performance Quadrajet produced on a large scale was the Performer RPM Q-jet from Edelbrock. The part number for the RPM Q-jet is 1910. These units featured a host of high-performance modifications, generous idle-fuel calibrations and were advertised as 850 cfm. The RPM Q-jet was only offered in a side fuel-inlet model. It is based on the late-style large primary bore Chevrolet castings and came with an electric choke to make it easy to retrofit to almost any application.

Here our RPM Q-jet is shown disassembled for inspection.

Rated at 850 cfm, the air flaps open to a full-open point of 1.270 inches as measured from the leading edge of the flaps back to the opening in the air horn as shown.

position. They would use the late-style "M" primary metering rods, small float and quick opening choke pull-off. The full open angle of the secondary air flaps was set at 1.270 inches, as measured from the leading edge of the flaps to the rear of opening in the casting.

The secondary air flaps also received two additional holes and two small tabs to help with air distribution into the secondary throttle bores.

The basic castings used were identical to the later Chevrolet side-inlet large primary-bore carburetors, using the late-style

The secondary air flaps contain round holes and "tabs." The round holes are located inboard and on the front half of the air flaps. They allow some fuel to be pulled into the engine from the accelerating holes located just above the air flaps. The "tabs" on the rear portion of the air flaps are designed to help air and fuel distribution into the carburetor at full throttle.

The Adjustable Part Throttle (APT) screw is slotted so it can be adjusted with a small flat-tip screwdriver.

A removable set screw is located in the air horn just above the APT screw. This allows easy access to the APT for quick, fine adjustments without taking the carburetor apart.

The RPM Q-jet is based on the later-style single main air-bleed design. The single pair of main air bleeds is located in the air horn and extend at an angle into the incoming air stream on the primary side of the carburetor.

As with most 1976 and later carburetors, the upper idle air bleeds are located in the main casting instead of the air horn. We have inserted two small drill bits into the upper idle air bleeds to show their location above the idle channel restrictions. They are drilled at a down angle into the main primary bores.

The RPM Q-jets were equipped with electric chokes. This makes them more universal and can be easily retrofitted onto earlier engines originally set up for divorced chokes. As with most late-model emission-years carburetors that used an electric choke, the choke held in place with rivets to prevent tampering. The rivets can be removed and the choke housing tapped for screws if adjustments are required.

Late-style Chevrolet linkage was used with additional holes drilled to provide flexibility in hooking up throttle linkage, return springs, and transmission downshift cables.

more universal. The standard late-style Chevrolet throttle linkage was used, with several provisions for installing throttle cable hook-ups and a lower location for return springs and a transmission downshift linkage.

As delivered from Edelbrock, the RPM Quadrajets had a very generous idle fuel calibration. This made them perfect for heavily cammed engines, but often somewhat rich for "mild" engines. We've recalibrated scores of them through the years, and several different idle calibrations could have been used. As with any other high-performance Quadrajet, best results will come from calibrating the carburetor exactly for the application.

Most units are somewhat rich at idle, lean at cruise, and rich at full throttle, depending on the applications they are being used on. We obtained a "virgin" unit and have recorded the following specifications for the calibration:

RPM Q-Jet Factory Calibration	
Idle discharge holes:	.085 inch
Idle bypass air:	.110 inch
Lower idle air bleeds:	.060 inch
Upper idle air bleeds:	.070 inch
Idle channel restrictions:	.055 inch
Idle tubes:	.040 inch
Main air bleeds:	.046 inch
Main jets:	.072 inch
Primary metering rods:	52M
Accelerating well	
holes (main casting):	.050 inch
holes (air horn):	.055 inch
Accelerator pump	
disharge holes:	.025 inch
Secondary air	
bleed tubes:	.020 inch
Secondary	
metering rods:	.041 inch (CE)
Hanger:	I

adjustable part throttle (APT) system and float arrangement.

They used a single pair of main air bleeds in the air horn, and the upper idle air bleeds were in the main casting just above the idle channel restrictions.

They came with an electric choke as standard equipment to make them

Complete tuning kits are available for the RPM and other Edelbrock Q-jets. They contain a wide variety of parts including an extra .149-inch needle/seat assembly. This complete Competition Calibration Kit is part number 1992. This kit will also work for many late-model factory Q-jets.

CALIBRATION KIT FOR MODEL M4M ('75 & LATER) AND EDELBROCK #1910 — KIT CONTAINS:

4 Primary Metering Rods (pairs)
Dual Taper for Q-Jet Model M4M ('75 & later) and Edelbrock #1910. Dimension indicates maximum diameter of taper.

(.048") ....#1941	(.052") ....#1945	
(.050") ....#1943	(.054") ....#1947	

5 Primary Metering Jets, all years (pairs)

(.072") ....#1972	(.075") ....#1975	(.077") ....#1977
(.073") ....#1973	(.076") ....#1976	

5 Secondary Metering Rods, all years (pairs)
Dimension indicates tip diameter. Letters are for identification purposes only.

CC (.0300") ....#1950	CK (.0527") ....#1952	CL (.0667") ....#1954
CE (.0410") ....#1951	AY (.0567") ....#1953	

Secondary Metering Rod Hangers, all years
Number in parenthesis indicates hanger height from rod hole to hanger mounting surface.

Hanger "B" (.520") ....#1960	Hanger "P" (.590") ....#1963
Hanger "G" (.545") ....#1961	Hanger "V" (.615") ....#1964
Hanger "K" (.565") ....#1962	

Power Piston Springs Assortment #1994

8" ....(bright yellow)	5" ....(orange)	
6" ....(black)	4" ....(golden yellow)	

Needle and Seat Assembly #1980
High capacity/high flow .149" diameter needle and seat assembly.

High-Performance Accelerator Pump Plunger and Spring #1982
Shorter than standard street plungers, this pump plunger and spring gives additional pump shot volume.

Edelbrock Corp., 2700 California St., Torrance, CA 90503 • Tech Line: 310-782-2900 • Not legal for sale or use on pollution controlled vehicles. © 1994 Edelbrock Corp. Rev. 1/94

To avoid guesswork, the #1992 Calibration Kit provides a specifications sheet. Notice that the primary metering rods are listed as "Dual Taper" and the number indicates the largest diameter of the tapered portion of the rod. These "M" rods feature .036-inch tips.

Edelbrock built and sold thousands of these carburetors until discontinuing production and sales in 2005. They show up frequently on Ebay and at swap meets, and make an excellent high-performance carburetor. Even though Edelbrock dropped their line of Q-jet carburetors, they continue to support them with a wide variety of high-performance parts, including jets, rods, power piston springs, accelerator pumps and tuning kits.

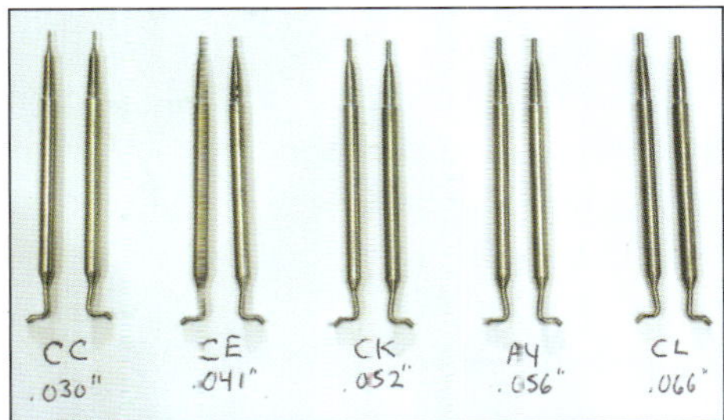

The #1992 kit contains five pairs of secondary metering rods. These metering rods fit all years and models of Q-jets and can be purchased separately.

Most of the items found in the RPM tuning kit can be purchased separately. Edelbrock also makes a tuning kit for early-style carburetors and offers a wide variety of primary jets and primary metering rods. Secondary metering rods are available in five different sizes:

CC:	.030 inch
CE:	.041 inch
CK:	.052 inch
AY:	.056 inch
CL:	.066 inch

The approach to custom tuning the Edelbrock RPM Q-jet is similar to any other late-style single main air-bleed APT carburetor. The secondary accelerating system exit holes are located over the secondary air flaps. The flaps can be notched or the holes relocated under the flaps to improve fuel delivery when the secondary throttle plates are opened.

If the RPM Q-Jet is to be used in a mild application, it may be advantageous to install slightly smaller idle tubes to provide full control of the idle mixtures. A good starting point for idle tubes is .036 inch for near-stock engines, or those with cams small enough to provide about 15 inches or more vacuum at idle speed. For more radical engines, the stock idle system is usually pretty close. In any case, it is best to test the carburetor as delivered, as several different idle-system calibrations were used through the years of production.

We have provided a basic recipe that will make a very good starting point for the RPM carburetor. The calibration listed has been street-, track-, and dyno-tested and will be very close for most high-performance applications:

The secondary acceleration holes exit over the air flaps as shown. Even though two large holes are present in the air flaps to allow the fuel to be drawn into the engine, when the secondary throttle plates first open, secondary performance can be improved by "notching" the air flaps directly in line with the holes. The holes can also be relocated under the flaps.

RPM Q-Jet Basic Recipe

Idle discharge holes:	.096 to .099 inch
Idle bypass air:	.099 to 110 inch*
Lower idle air bleeds:	.070 inch
Upper idle air bleeds:	.070 inch
Idle channel restrictions:	.055 to .059 inch
Idle tubes:	.038 to .040 inch**
Main air bleeds:	.080 to .086 inch
Main jets:	.076 to .077 inch
Primary metering rods:	54M
Accelerating well holes (main casting):	.050 inch
Accelerating discharge holes (air horn):	.055 to .059 inch
Accelerator pump disharge holes:	.028 to .032 inch
Secondary air bleed tubes:	.036 inch
Secondary metering rods:	.041 inch (CE)
Hanger:	P

** Engines with smooth idle characteristics may require less idle by-pass air. The by-pass air holes in the base plate can be carefully plugged with a small piece of aluminum or brass rod, then re-drilled to a smaller size, if needed.*

*** Smaller idle tubes may be beneficial for near-stock engines or those that have a lot of vacuum at idle speed.*

While the carburetor is apart, set the float to exactly ¼ inch. Test the release time for the choke pull-off; it should release in less than two seconds. If not, modify the unit or obtain a quicker pull-off. Adjust the secondary air flap spring to ½ turn as a starting point.

As with tuning any Q-jet, the power piston spring should be as strong as possible, yet stay down at idle. This provides the best throttle response with quick throttle opening. With the RPM Q-jet, as with many other Q-jets, we can observe the position of the power piston with the engine running at idle speed. With the RPM Q-jet, we can remove the APT access screw and look at the position of the steel post that contacts the screw to see if it is down at idle speed.

After recalibration and before placing the carburetor in service, set the APT adjustment screw about one to two turns counterclockwise from seated. Leave the set screw out of the air horn and fine tune the APT once the carburetor is placed in service.

Quadrajet Secondary Metering Rods

Degrees of Secondary Air Valve Opening vs. Rod Diameter in Secondary Jet
Rod diameter measured in ten thousandths of an inch. (.0001")
Secondary Jets are all .1360" and non-removable from carburetor

I.D. Code	0 deg.	20 deg.	40 deg.	60 deg.	70 deg.	80 deg.	90 deg.	Tip length	GM Part #
AD	1346	1280	1084	812	627	397	222	S	7033772
CB	1347	1287	1197	997	897	547	297	S	7042335
BV	1347	1287	1197	1097	897	547	297	S	7040724
DX	1320	1247	1090	600	400	300	300	M	1706471
DC	1347	1236	1129	1024	599	300	300	M	7047816
CC	1347	1236	1129	1024	599	300	300	M	7042356
BY	1287	1217	1062	817	667	317	317	M	7040856
DU	1280	1244	1075	847	547	337	337	M	1705995
DS	1310	1244	1075	847	547	337	337	M	1705661
DF	1339	1262	1075	847	547	337	337	M	7048512
DG	1342	1244	1075	847	547	337	337	M	7048890
CF	1347	1244	1075	847	547	337	337	M	7044775
DM	1334	1281	1155	853	655	482	393	S	1705022
CJ	1339	1244	1075	807	419	419	397	L	7045780
AX	1339	1244	1075	807	419	419	397	L	7033549
BB	1339	1244	1075	807	419	419	397	L	7034335
BW	1317	1247	1087	897	597	397	397	M	7040767
CA	1322	1277	1147	937	797	397	397	M	7042304
BH	1342	1290	1159	932	519	397	397	M	7035916
BM	1342	1290	1159	932	519	397	397	M	7037744
CM	1342	1294	1165	932	519	397	397	M	7045840
BG	1352	1294	1165	932	519	397	397	M	7034822
BF	1325	1315	1192	1021	891	639	397	S	7034400
BK	1325	1315	1193	1021	891	639	397	S	7037295
BJ	1333	1315	1193	1021	891	639	397	S	7036077
CS	1330	1287	1197	1097	897	687	397	S	7045924
BP	1337	1287	1197	1097	997	687	397	S	7038034
CE	1347	1244	1075	847	547	410	410	M	7043771
BN	1329	1288	1147	957	793	503	410	S	7036671
BL	1329	1311	1188	1021	891	691	410	S	7037733
BE	1329	1311	1188	1021	891	691	410	S	7034377

I.D. Code	0 deg.	20 deg.	40 deg.	60 deg.	70 deg.	80 deg.	90 deg.	Tip length	GM Part #
DA	1333	1244	1075	847	577	440	440	M	7046010
CY	1333	1244	1075	847	577	440	440	M	7046004
CV	1332	1291	1154	927	527	527	527	L	7045984
AU	1345	1291	1154	927	527	527	527	L	7033655
CK	1345	1291	1154	927	527	527	527	L	7045781
AH	1345	1291	1154	927	727	527	527	M	7033812
EJ	1355	1350	1200	1047	837	687	540	S	1708193
EA	1339	1262	1172	1000	825	721	547	S	1708009
CR	1339	1262	1172	1097	997	747	547	S	7045923
BU	1347	1262	1172	1097	997	747	547	S	7040725
CP	1320	1247	1090	879	743	578	567	M	7045842
DH	1332	1247	1090	879	743	578	567	M	7048992
ED	1332	1247	1090	879	743	578	567	M	1708187
BA	1332	1247	1090	879	743	578	567	M	7034337
AL	1349	1247	1090	879	743	578	567	M	7033680
CD	1351	1264	1107	755	567	567	567	L	7042719
DZ	1325	1290	1150	879	743	578	567	M	1707526
AV	1338	1292	1132	907	757	567	567	M	7033182
AP	1345	1292	1132	907	757	567	567	M	7033981
AJ	1345	1292	1132	907	757	567	567	M	7033628
AZ	1350	1292	1157	957	567	567	567	L	7033889
EG	1350	1292	1157	957	847	697	567	S	1708188
AK	1350	1292	1157	957	847	697	567	S	7033104
CH	1350	1292	1157	957	847	697	567	S	7045779
CX	1346	1294	1165	932	567	567	567	L	7045985
BZ	1352	1294	1165	932	567	567	567	L	7042300
AY	1352	1294	1165	932	567	567	567	L	7033830
EK	1340	1294	1165	932	790	567	567	M	1708240
EF	1340	1294	1165	932	790	617	567	M/S	1708188
CN	1340	1294	1165	932	790	617	567	M/S	7045341
AR	1352	1294	1165	932	790	617	567	M/S	7033171
EE	1323	1250	1093	882	746	581	570	M	1708187
DR	1323	1250	1093	882	746	581	570	M	1705365
DV	1348	1294	1168	935	793	620	570	M/S	1706217
BD	1332	1312	1168	939	805	577	577	M	7034365
BC	1352	1294	1174	1015	901	599	581	M	7034300
DN	1355	1297	1108	1019	904	602	584	M	1705370
BT	1338	1292	1132	907	787	597	597	M	7040601

I.D. Code	0 deg.	20 deg.	40 deg.	60 deg.	70 deg.	80 deg.	90 deg.	Tip length	GM Part #
EC	1350	1292	1157	847	697	610	610	M	1708109
DJ	1337	1262	1172	1097	997	847	747	unique rod-taper continues down to 647	7048908
CL	1350	1294	1176	984	667	667	667	L	7045782
AT	1350	1294	1176	984	667	667	667	L	7033658
DL	1336	1260	1127	939	824	711	679	M/S	7048392
EH	1325	1267	1137	945	829	721	686	M/S	1708188
DP	1325	1267	1137	945	829	721	686	M/S	1705353
DB	1334	1294	1177	1010	907	797	697	S	7047306
BX	1352	1294	1177	1010	907	797	697	S	7040797
AN	1352	1294	1177	1010	907	797	697	S	7034320
DT	1337	1265	1180	1100	1000	700	700	M	1705665
CT	1337	1294	1176	984	849	774	774	M	7045983
AS	1350	1294	1176	984	849	774	774	M	7038256
CG	1350	1294	1176	984	849	774	774	M	7045778
DW	1334	1294	1170	1010	883	874	874	L	1706243
DY	1337	1305	1191	1010	883	874	874	L	1707426
DE	1337	1305	1191	1010	883	874	874	L	7048092
BR	1350	1294	1176	984	897	897	897	L	7038910
AW	1351	1283	1209	1036	965	965	905	2-step tip	7033194
CZ	1345	1292	1157	947	947	947	947	LL	7045936
EB	1345	1292	1157	947	947	947	947	LL	1708046
BS	1350	1292	1157	947	947	947	947	LL	7038911
DK	1345	1298	1169	997	997	997	997	LL	7048919
DD	1345	1298	1169	1047	1047	1047	1047	LL	7043091

All dimensions are in 1/10,000 of an inch

Tip Length Legend:

LL: Extra long power tips supply richest mixture starting at 60° of air valve opening. Never use in performace applications using greater than 70° air valve opening.

L: Power tip starts at 70° air valve opening. Tip considered to begin at that part of the rod that is within .003" of rod's minimum diameter

M: Power tip starts at 80° air valve opening. Tip considered to begin at that part of the rod that is within .003" of rod's minimum diameter

M/S: Power tip starts at 80° air valve opening. Tip considered to begin at that part of the rod that is within .005" of rod's minimum diameter.

S: Power tip starts at 90° air valve opening.

(This information was generously provided by Damon Nickle.....Much thanks!)

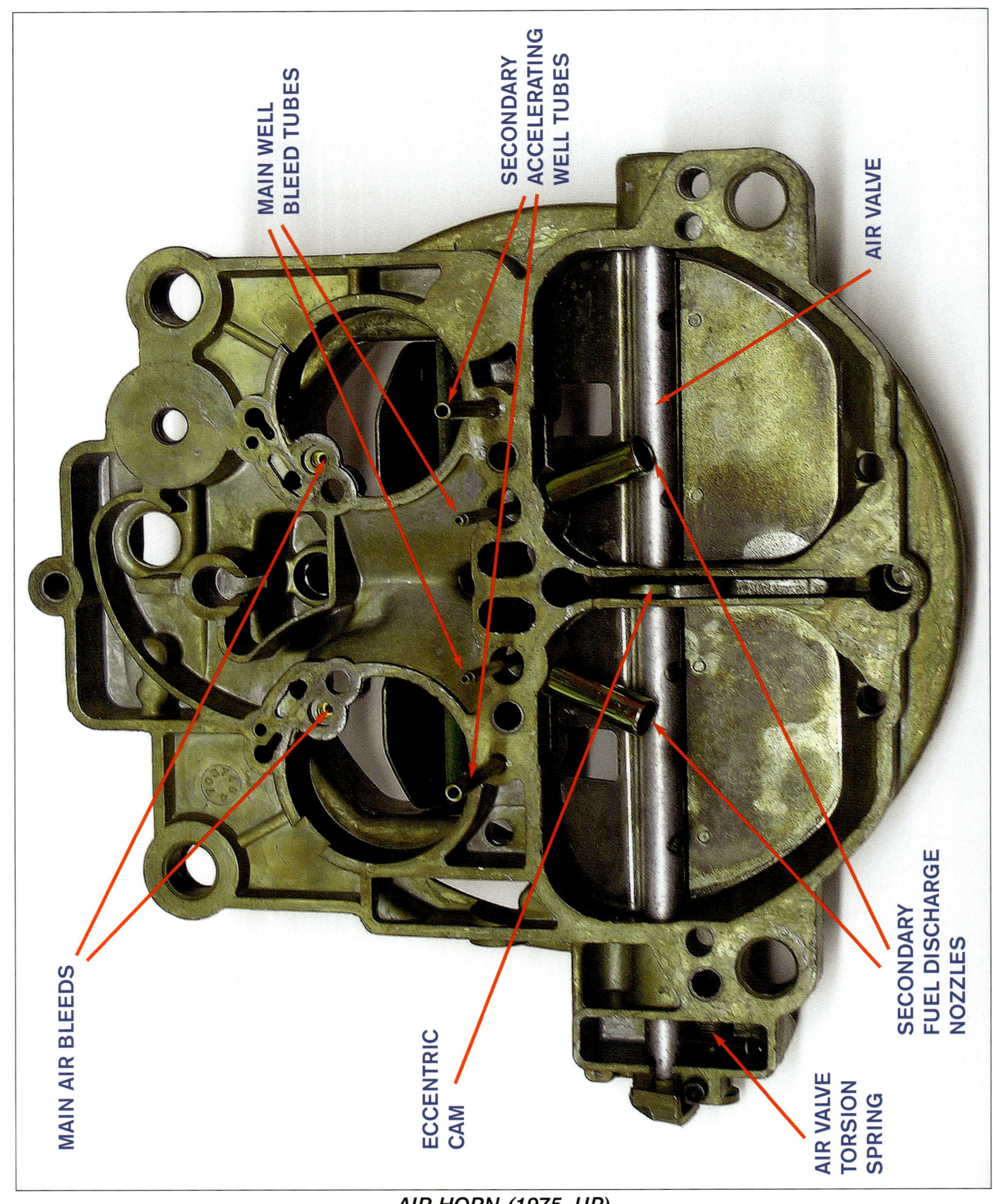

AIR HORN (1975–UP)

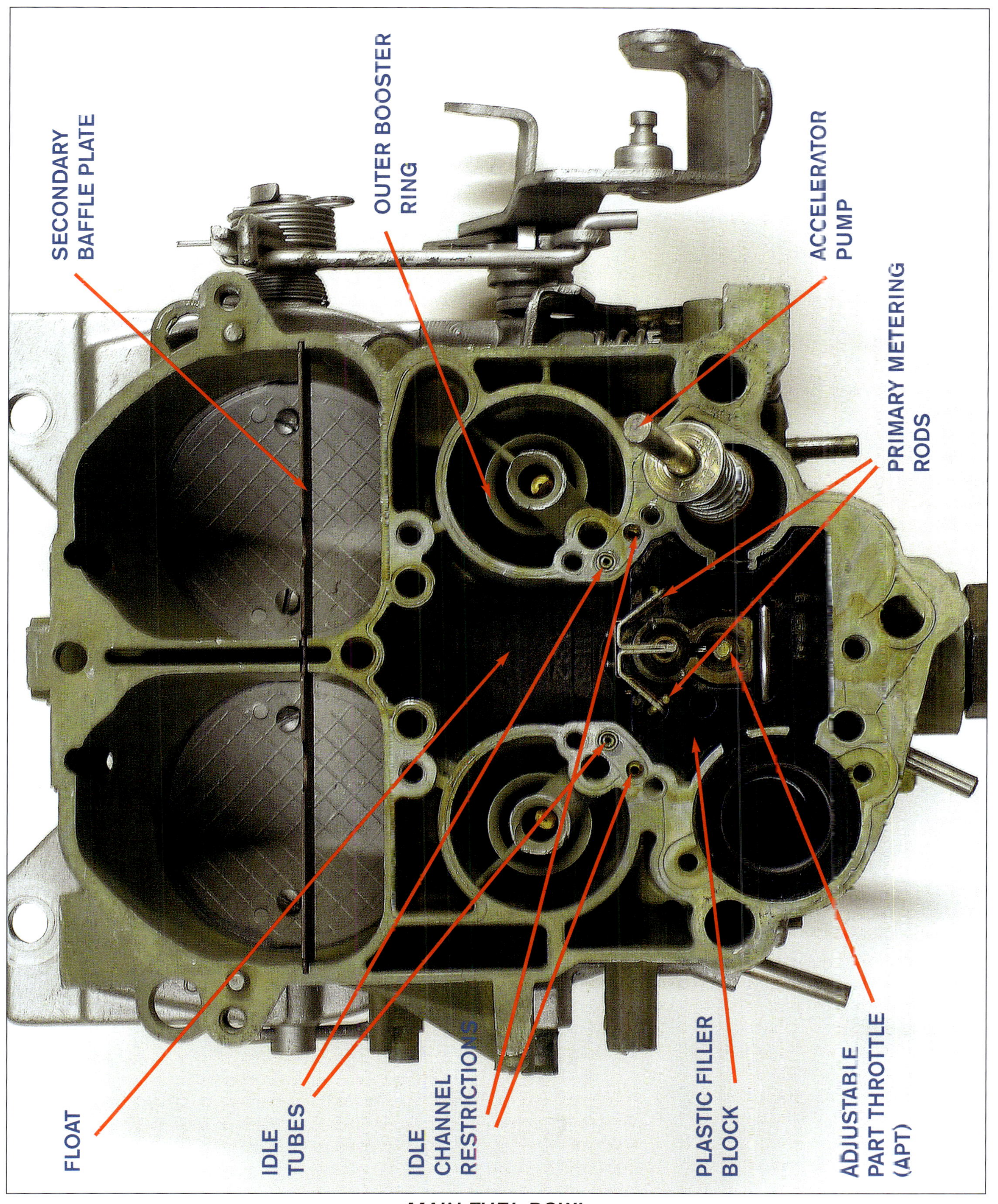

MAIN FUEL BOWL

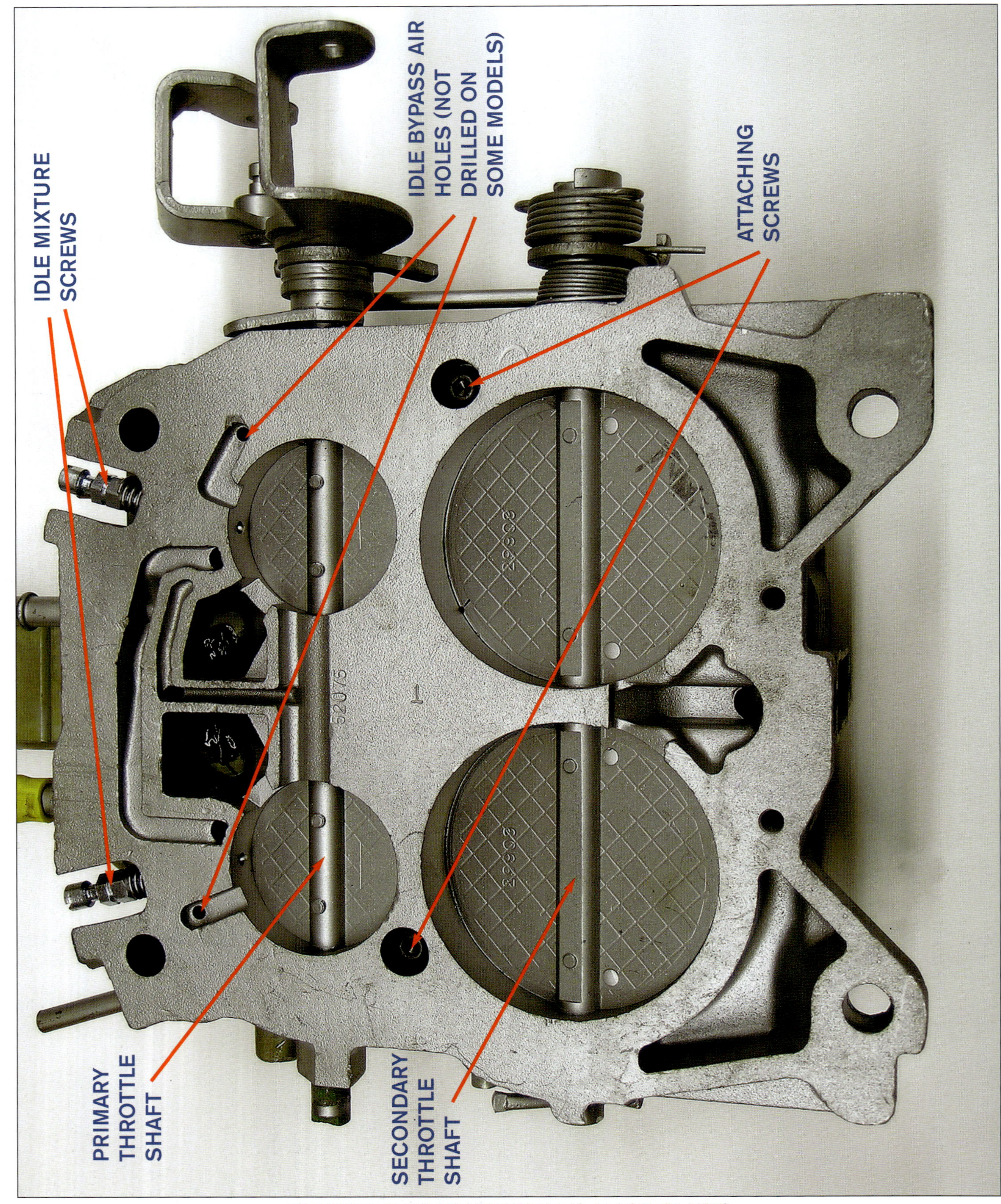

THROTTLE BODY ASSEMBLY (BASE PLATE)